PRAISE FOR *IT IS POSSIBLE*

* * *

The most intelligent, comprehensive, and compelling argument ever advanced against nuclear weapons.
GEORGE LEE BUTLER
U.S. AIR FORCE (RET.), FORMER COMMANDER, STRATCOM

Ward Wilson's book makes me believe that the eradication of nuclear weapons is feasible in our lifetime.
OSCAR ARIAS
FORMER PRESIDENT OF COSTA RICA
NOBEL PEACE PRIZE LAUREATE

Provides the inspiration people need to eliminate these weapons.
BEATRICE FIHN
FORMER EXECUTIVE DIRECTOR OF ICAN
NOBEL PEACE PRIZE LAUREATE

Arguably the most important contribution to the debate over the efficacy/fallacy of nuclear deterrence ever written.
MARTIN SHERWIN
PULITZER PRIZE-WINNING HISTORIAN OF NUCLEAR WEAPONS

You owe it to yourself to read this remarkable message of hope.
JOE MORRIS DOSS
EPISCOPAL BISHOP (RET.)

Both practical and wise.
FREEMAN DYSON
PHYSICIST, INSTITUTE FOR ADVANCED STUDY

The world doesn't need nuclear weapons and this book proves this fact clearly and firmly.
DR. SHIRIN EBADI
FOUNDER OF THE DEFENDERS OF HUMAN RIGHTS CENTER
NOBEL PEACE PRIZE LAUREATE

This highly readable book debunks the myth that nuclear weapons can't be eliminated and shows clearly that, indeed, It Is Possible.
IRA HELFAND
PAST PRESIDENT OF IPPNW
NOBEL PEACE PRIZE LAUREATE

A book that should be read by every decision-maker and everyone else across the world. Together we can end the insanity of nuclear weapons.
J. RAMOS-HORTA
PRESIDENT OF TIMOR-LESTE
NOBEL PEACE PRIZE LAUREATE

Ward Wilson is the most innovative thinker about nuclear weapons anywhere. It Is Possible is truly a masterpiece. Its comprehensive analysis shows not only the uselessness of nuclear weapons but outlines new ways of thinking about them that make their elimination not only feasible but the only practical course for the survival of humanity. A fascinating read, it is well worth the time of anyone concerned with the planet's future.
BARRY BLECHMAN
CO-FOUNDER OF THE STIMSON CENTER

Ward Wilson's innovative proposals should be pondered by all those who wish a world free of nuclear weapons.
SERGIO DUARTE
PUGWASH CONFERENCES
NOBEL PEACE PRIZE LAUREATE

This is a factual book, a book that draws on history, and that makes pragmatic arguments. If you are interested in a counterpoint to conventional thinking, if you want to explore new realist arguments about nuclear weapons, this is the book for you.
BERNARD NORLAIN
GÉNÉRAL D'ARMÉE AÉRIENNE(2S) (RET.)
FORMER COMMANDER OF THE CONVENTIONAL AIR FORCES OF FRANCE
PRESIDENT OF NUCLEAR DISARMAMENT INITIATIVE

Nuclear weapons can be eliminated. Ward Wilson tells us how.
JODY WILLIAMS
FOUNDING COORDINATOR OF ICBL
NOBEL PEACE PRIZE LAUREATE

Ward Wilson is known in his own country, in my country, and in nations across the world as a man of intellect and honour, who is dedicated to educating on behalf of a World Free of Nuclear Weapons. I am proud to endorse this important new book and the vital message that it carries for us all.
BILL KIDD
MEMBER OF THE SCOTTISH PARLIAMENT

A superb book, one that will make you think about nuclear weapons in ways you had not thought about before. It Is Possible is a must-read for both policymakers and citizens alike.
DR. ROLF MÜTZENICH
CHAIRMAN OF THE SPD PARLIAMENTARY GROUP, GERMAN BUNDESTAG

A breath of fresh air. At a time in my career when everything in the nuclear weapons debate seems stale and shopworn, I found It Is Possible full of original ideas, previously unused historical examples, and guided by an entirely new approach. I really like the way it takes realists at their word and then points out that a realist view of human nature would say that wars and mistakes will happen—which makes it all the more important to get rid of nukes. And I love the critique of the 'disinvention' bull#@^t.
DR. DAVID P. BARASH
PROFESSOR OF PSYCHOLOGY (EMERITUS), WASHINGTON UNIVERSITY

In this stunning, breakthrough work, Ward Wilson brilliantly dismantles the false claims about nuclear weapons that have kept a nuclear sword of Damocles dangling over our heads for so long.
RICHARD RHODES
PULITZER PRIZE WINNING HISTORIAN OF NUCLEAR WEAPONS

If you only ever read one book about nuclear weapons, let it be this one. Easy to read, meticulously well-reasoned, it has an almost disarmingly straightforward answer for every conceivable challenge to the idea that a future free from nuclear weapons can exist. This guidebook for eliminating nuclear weapons will give even the most resolute cynic reason to hope that it is, indeed, possible.
EMMA PIKE
NUCLEAR DISARMAMENT CONSULTANT AND ACTIVIST

IT IS POSSIBLE
A Future Without Nuclear Weapons

WARD HAYES WILSON

AVENUES THE WORLD SCHOOL PRESS

AVENUES THE WORLD SCHOOL PRESS

11 Madison Square North
16th Floor
New York, NY
10010-1420 USA
https://press.avenues.org

Copyright © 2023 by Ward Hayes Wilson
All rights reserved.

Avenues supports the right to free speech and the value of copyright. The purpose of copyright is to encourage writers and artists to produce works that enrich our culture. Thank you for supporting them.

No part of this book may be reproduced in any form without written permission of the copyright owner.

ISBN: 979-8-9850518-4-1

This book is dedicated to
Freeman Dyson, Martin Sherwin, and Barry Blechman:
teachers, guides, and friends.

CONTENTS

	ACKNOWLEDGMENTS	i
	INTRODUCTION	1
1	THE GAME WAS RIGGED	21
2	SOCIETY-WIDE MISTAKES	37
3	IN THE BEGINNING	49
4	REFRAMING IT ALL	65
5	BIGNESS—IT'S OVERRATED	81
6	THEY WEREN'T SHOCKED	93
7	LUMPS OF COAL	113
8	CRUELTY AND WAR	143
9	SYMBOLS	153
10	DETERRENCE THEORY	173
11	DETERRENCE REALITY	189
12	ELIMINATION	209
	APPENDIX	239
	INDEX	248
	ABOUT THE AUTHOR	250

ACKNOWLEDGMENTS

This book was originally intended to be a compilation of essays—some published, some unpublished—that I had written over the previous six years. However, once I got to the actual writing and tried to give the argument a stronger narrative flow, the essays somehow kept getting revised. The revisions got so out of hand that none of the original published articles now has much of its original shape. However, a careful reader can still see the outlines of those essays in several chapters and some sections of text drawn directly from those essays.

I am grateful to the editors of the *Bulletin of Atomic Scientists*, the European Leadership Network, and Inkstick for permission to use the articles that were published in their pages.

"How nuclear realists falsely frame the nuclear weapons debate" appeared in the *Bulletin of the Atomic Scientists* (online edition) on May 7, 2015. It served as the basis for portions of chapters 3 and 11. "Reconsidering nuclear deterrence" appeared on the European Leadership Network's website on March 1, 2022. Parts of it were used in the chapters on nuclear deterrence, chapters 10 and 11. "A World Without Nuclear Weapons: Pipe Dream? Or Inevitability?" was published by *Inkstick* on February 4, 2020. Large chunks of it were used in chapter 4. "How to Eliminate Nuclear Weapons" was published as a four-part series in *Inkstick* on March 29, April 2, April 5, and April 9, 2021. Ideas and arguments from it were used throughout this book. I am especially grateful to *Inkstick*, a publication with a fresh outlook that is willing to publish new and unconventional ideas in a field where innovative thinking and new ideas are rarely rewarded and sometimes outright punished.

This book would not have been possible without the advice and encouragement of the board members of RealistRevolt. William Ranney, Kevin Ellis, and Paul Ingram all made substantial contributions of time, advice, and expertise. Without them this work could not have emerged in the shape that it did. I have come to lean on their experience and advice and hope to continue to do so for many years to come.

I am indebted to the people of Norway, whose Foreign Ministry awarded me the grant that made it possible for me to travel for six years and meet and talk with diplomats, scholars, and ordinary people in so many countries. It also made possible periods of dedicated thinking that deepened my understanding, that gave me the space to step back far enough to see the framings that had distorted thinking on this important matter for so long, and that, in many ways, launched me into a serious pursuit of the elimination of nuclear weapons.

My debts are many, and my gratitude must be poured out widely. I have been hosted by people who were willing to let a stranger talking about a frightening subject eat at their table and sleep in their home. I have been invited by people to talk to the audiences that they had assembled and welcomed to gatherings and conferences by people willing to hear something new. To all of them I send my humble thanks and deep gratitude. They are the rare breed of people who take concrete steps to make this a safer, better world.

I am grateful to Avenues The World School for publishing this book and their support throughout the process. Special thanks to William Lidwell, who oversaw the project and who was willing to take risks to explore new approaches to solving this problem, and to Rebecca Strauss, who managed the editorial process and who was both encouraging and wise. I am also indebted to David Umla whose tireless meticulousness strengthened the final result immensely.

I am also grateful to the members and leadership of Winnetka Presbyterian Church. No one should attempt a large and difficult task without the support of some sort of spiritual community. Community is essential whenever one is pursuing a formidable undertaking, and spiritual community strengthens us to face the darkest and harshest parts of human reality. (Perhaps it is even more important in a country like mine where we so often focus on material objects of status and comfort.)

And finally, my greatest debt is to A.L.T., without whom this book would not have been possible, and to whom I owe a debt that I cannot hope to repay.

INTRODUCTION

A small child, maybe five or six years old, is in the middle of a large front yard looking up. It is windy and cold—it is fall. He stands alone, unnaturally still. He stares at the clouds that hang threateningly in the sky.

What is he looking at? He's wearing a coat, but the wind is gusting, and standing there he must feel the chill. Still, he stares, unmoving—watching and watching. A small sentinel in a large yard somewhere in central New Jersey. A leaf blows past his feet.

A child's understanding of the world is uncertain. They do not fully understand how things work. Sometimes children make mistakes about what they have power over and what they don't. They believe, wrongly, that they are responsible for things that are actually beyond their control. They imagine, for example, they could have prevented the car accident, the divorce, or some other life-altering, chance occurrence. Maybe this child is looking skyward because he believes he can make some difference in the course of events.

Why is he standing and staring so intently on a cold October day?

If we could see inside his heart, we would see that he is in the grip of a fear unlike anything he has ever known.

* * *

I was a sunny, smiling child. Even at six, when my parents had a party, they would bring me out—the living room filled with towering grown-ups—and get me to tell a sweet little joke. I loved it when the adults, surprised and delighted, burst out in laughter.

IT IS POSSIBLE

My childhood consisted of running through fields, crawling around brambles, digging busily, and chasing in my pajamas across the lawn on summer evenings after fireflies. Then, one week, I realized that something was wrong. Seriously wrong. It felt like a cold breath on my neck. My parents seemed suddenly strange. Children have extremely sensitive emotional antennae. But like the first warning signs of danger in a sci-fi movie, I couldn't figure out what I was sensing. The operator takes the bulky headphones off, a troubled look on his face, and turning to the captain leaning on the console—waiting, expectant—he says, confused, "I don't know what it is, Captain. I can't make it out. It's unlike anything I've ever heard before." Those signals filled me with unease. What was making my parents behave so differently—so solemn and stern? Something was happening. I was afraid.

I watched them and catalogued the signs. My parents weren't smiling. They didn't laugh. The muscles around their mouths and jaws seemed tight. (Was there a disease that started with tightness around your jaw?) They were looking at each other sometimes, sending the unspoken, grown-up messages that we—my sister and I—could almost never decipher. And all the time, as my family worked its way through life's everyday activities, the eerie, unfamiliar signals kept repeating and repeating. They seemed to be getting stronger.

Finally, I asked my mother. "What's going on? What's wrong?" And, looking back, I can't really fathom what she did next. She told me the truth.

Without preamble I found myself in a world with forces so far beyond my control I had never even suspected they existed. She told me "bad men" had put missiles ninety miles from the coast of the United States. The missiles had bombs on them that could blow up "everything." And they might come down out of the sky at any time.

It was the Cuban Missile Crisis, and what my mother was explaining, in words a six-year-old could understand, was an international crisis that threatened much of the world as well as threatening to tear apart everything I knew. Described as the week that the world held its breath, for me it seemed to last much, much longer than a week. It was cold, and I was frightened.

I can remember the wind and the gray clouds that filled the sky. What was I watching for, you may wonder? What kept me standing there in the front yard in the cold? I was watching for the first small specks in the sky, the telltale sign of the rockets coming in, that would bring the end of my world, that would be the death of my mother and father, my sister, our dog Kelly, and me.

INTRODUCTION

For many years, if you asked the average person in the United States what they thought about nuclear weapons, they would say, "I never think about them." And that was probably true. They didn't lie in bed and worry. They didn't talk in an undertone with their closest friends. They didn't stand at the bus stop and suddenly turn away as an image flashed unexpectedly into their mind's eye of fire and a column of smoke rising over their city.

But nuclear weapons are like long-ago family strife. They're like the submerged memory of a terrible accident. They have burrowed down deep into our psyches so that no sign of their existence shows on the surface. People don't talk or even consciously think about them. But they are there, leaking poison into our veins, influencing who we are.

The International Red Cross polled millennials worldwide in 2019 and asked them, among other things, if they thought a nuclear weapon would be detonated somewhere in the world in the next ten years. The figures for the United States were startling. Fifty-eight percent said yes. Asked if they thought there would be a war on the scale of World War II in their lifetimes, the same percentage again said yes. And this fear about nuclear war existed before the current invasion of Ukraine and President Vladimir Putin's repeated threats to use nuclear weapons.

Most people in the United States believe that if a single nuclear weapon gets used, all of them will get used. It's a widespread assumption and arguably correct. So, think about those two facts. Some fairly large percentage of Americans is carrying around the belief in their heads that a nuclear weapon will be used in the ten years following 2019 and that probably it won't be just one weapon but, rather, all of them. This will happen before they finish paying off the mortgage, before Jane finishes college, before Nana dies. They are carrying around the belief that their world will be catastrophically transformed by destruction and fire. There are millions of people who hold this dark belief somewhere in their heads … without ever talking about it.

How do you live with the expectation of a life-altering disaster and never acknowledge it? Or claim you never think about it? Nuclear weapons loom large in our lives, apparently, but we resolutely pretend not to see them.

It's hard to think about nuclear weapons, in part because our feelings about them are so strong. Feelings, tucked away and ignored, but still beating with a malignant insistence. Seeing the huge mushroom cloud, we feel insignificant. Looking back over the more than seventy-five years of failure to get rid of them, we feel beaten. Watching the slow expansion of the number of countries that have them, we feel out of control. The weapons remind us of Hiroshima, the stain that makes conservatives bristle and shout, "It was justified!" and that makes liberals feel guiltier still—and which neither conservatives nor liberals can seem to let go of. (There is still today a thriving debate over whether the bombings were justified. Think of it: three-quarters of a century later, and we still can't make peace with ourselves about that decision.)

As we live through our days, worrying about money, disliking our jobs, putting tempting things into our mouths that we know have too much sugar, and only ever distracted when we are passively watching, watching (always watching)—all the time nuclear weapons are there, somewhere, unseen in the deep recesses of our minds. We fear them, but we cannot seem to be rid of them.

When something forces us to turn and look nuclear weapons in the face—some young person with a clipboard and a petition or some world leader waving them as a threat—we feel angry. We don't want to think about nuclear weapons. We don't. Because we are so helpless. What can we do?

I understand the taboo on talking (or even thinking about) nuclear weapons. If there is a risk, you have to run—like a giant asteroid sailing out of deep space to collide with the Earth and destroy us all—and there's nothing you can do about it. The best thing you can do is put it out of your mind. What's the point of worrying about something beyond your control?

But I am concerned that by choosing to try to put the danger out of our minds, we are slowly wrapping ourselves in pessimism and passivity as a kind of protective shell. Sometimes, when I look at the troubles we have in the United States, I am reminded of prisoners on death row: people who sit and stare, trying not to hope, trying to let time pass while the most important parts of their souls slowly atrophy.

*   *   *

In this book, I will argue that something you (probably) think is impossible is, in fact, possible. I will walk you step-by-step through a realistic, factual, complete solution to eliminating nuclear weapons. The book is made up of a series of essays, some of which have previously been published and some of which are new. Each essay addresses one of the main objections that ordinary people have to getting rid of these weapons. Experts, of course, have their own set of arguments; but those arguments won't be addressed in this book. Experts have dominated this debate for far too long. There are no serious policy issues—issues that might directly involve the deaths of millions of people and that might even determine your death—that you are excluded from. You have a right to an opinion about these matters. Expertise is valuable and should be respected. But when survival is at stake, all of us have a right to be heard. This is especially true in the United States where we claim to hold democracy dear. The theory of democracy is that the best policy is the result of the collective wisdom of everyone, not just a select few.

So, these essays are for the rest of us, not for experts. They are based on common sense, on facts, and they strive to be entirely realistic. Sometimes this makes them somewhat harsh sounding, but when the issue at hand is the most dangerous threat there is to civilization, facing uncomfortable realities is required.

I'm not saying you'll find appalling descriptions of the effects of nuclear weapons in these pages. You won't. I don't think inciting horror in people is a useful approach to solving problems. I think if we're going to eliminate nuclear weapons, it's going to be because a realistic assessment of the facts shows conclusively that they're not worth keeping. I want to focus on facts, not stir up horror or moral indignation. There are no fairy tales about the goodness of human beings here, and there is no glossing over some of the unpleasant things that human beings have done in the past—and could well do again in the future. So, if you're not willing to look at some of the ugly things in human nature, close the book now.

* * *

Ironically, my journey with nuclear weapons was impelled by moral passion. In 1979, I was inspired by a speech made by Dr. Helen Caldicott, an Australian physician who talked movingly about the horrible medical consequences of nuclear explosions. Her words moved me to my core. Something resonated. After hearing Dr. Caldicott, I started reading, underlining, thinking, and pacing back and forth in my parent's living room when everyone else was asleep.[1] There had to be a way to eliminate nuclear weapons, I thought. They were too destructive and too dangerous to be safe.

My first impulse was to invent a new, stronger, and more inspiring moral argument against them. True, nuclear weapons were so powerful and so obviously decisive militarily that any argument against them would be an uphill battle. But the immorality of nuclear weapons was so obvious—it was the weapons' most apparent flaw—that using morality seemed to be the only way to eventually eliminate them. My parents happened to live near one of the best libraries on the east coast, Princeton University's Firestone Library, and at the time, anyone could wander and read in the stacks. So, I read a lot of Just War Theory and spent my days feverish with excitement. I was full of confidence. Bertrand Russell tried to argue against nuclear weapons using moral arguments. He'd done a fair job, but I would discover the line of argument that would really convince people.

But after almost a year, I realized there was a counterargument I had no answer to. I tried and tried to find a way around it, but there wasn't any. Suddenly, things looked dark. The obstacle I'd crashed into was the fact that necessity almost always trumps morality.

I couldn't escape the fact that in a crisis, when survival was on the line, people sometimes broke moral rules. Even long-standing and powerful moral rules or taboos didn't always hold. In a moment of mortal fear, when crisis clouded their thinking, when push came to shove, people's moral values sometimes failed. It seemed like a rule of human nature: When their lives are at stake, people often choose survival, even if it means using terrible and immoral means.

I'm not saying that morality is powerless. When the Romans insisted that they deny their faith, thousands of Christians went unresisting to the lions rather than surrender their beliefs; and the moral force of their example eventually (in a sense) conquered Rome.[2] Morality has had enormous power throughout human history, and any political theory that ignores that reality does not reflect the world as it actually is. I believe morality probably would work (sometimes) to prevent the use of nuclear weapons. But human beings are too inconstant; their attachment to their values fluctuates too much for morality to be enough to always prevent the use of nuclear weapons. If you rely exclusively on moral arguments, a day will come when someone faces a crisis, anger or fear will overwhelm their sense of right and wrong, and nuclear weapons will be used.

    I was crushed. My year of reading and thinking about the moral dimensions of nuclear weapons had been wasted. My earlier confidence suddenly seemed ridiculous. But if moral arguments couldn't prevent states from keeping and (I was sure) eventually using nuclear weapons, what could? If not moral arguments, then what? For a time, I despaired.

* * *

My mother died in 2017. I have tried to put myself into the shoes of the thirty-three-year-old woman she was during the Cuban Missile Crisis, with two small children, a dog, and a husband who commuted an hour into New York City every day. If a nuclear war had come, what was she planning to do? Suddenly the power goes out; perhaps there is a distant roar of explosions in Philadelphia. Would she put us into the car and head toward a

INTRODUCTION

devastated New York City to try to find her husband? Would she assume he was dead, try to accept the reality that she would never see him again, and take off in search of someplace safe, like, I don't know, Canada? Would we live in the basement, eating up the last of the groceries, trying not to breathe radiation, hoping that somehow help would come? And what was she obliged to say? With the possibility of nuclear war looming, should you tell your children about the danger? Do children have a right to know that they might die suddenly and unexpectedly?

In the fall of 1962, in the shadow of the most dangerous nuclear weapons crisis the world has ever known (so far), my mother decided to tell me the truth; and it changed me. I understood at a fundamental level how nuclear weapons threaten families and children, livelihoods, and ways of life. A nuclear war could shatter things that civilization has slowly and painfully constructed over thousands of years. That reality worked on me, ate at me.

*　　*　　*

Somehow, I got the idea, just after college, living in Washington, D.C., to walk fifty miles in a day. I had read that Robert Kennedy did it in 1962, and a friend of mine and I decided to try it in February 1981. We went in the dead of winter because that was when Kennedy went. It was a little crazy. But since we were going anyway, we contacted the Robert F. Kennedy Memorial Foundation, which was in D.C., and volunteered to raise a little money for them.

They found some people to sponsor us, and one morning, we set out at 5:00 am to walk along the C&O Canal towpath next to the Potomac—the same route Kennedy took. Kennedy had walked out from Washington toward Camp David. We got a friend to drive us out an hour from the city so we could walk back toward D.C. It was cold and dark. I remember the moon setting on our right, beautiful and reflecting on the river. Later the sun came up behind clouds to our left. And we walked. And walked. We told every story we knew. We sang songs. The hours passed. The sun eventually set—also over the Potomac. And then the moon, the same planetary object that we had seen disappear over the right-hand edge of the world, rose up behind the bare trees on our left. I knew from schoolbooks that the moon circled the Earth. But somehow to see it happen, to be confronted with the reality that while we had been walking the moon had traveled all the way around our world, astounded me.

At one point late in the afternoon, dirty, incredibly tired, flopped on the ground resting and thinking about the improbability of what we were doing,

I laughed and said to Rick, "You know, if we can do this, we can do anything."

It took us nineteen hours and forty-five minutes. As we stumbled into Georgetown (the fancy Washington neighborhood where the canal towpath ends), we felt triumphant, and half-expected to be greeted by crowds and a band. But all the well-heeled couples out for drinks or a late-evening meal saw were two dirty young men with swollen hands who looked like they'd been sleeping on the ground. Walking side by side, exhausted, we chuckled with amusement as couple after well-dressed couple gave us a wide berth as we passed.

Later David Hackett, then the head of the Robert F. Kennedy Memorial Foundation, asked me, "Ward, what do you care about?" Despite my failure to find a moral argument against them, I blurted out, "Nuclear weapons." David made me a Robert Kennedy Fellow, gave me a small stipend for books (I continued working as a temp at the Washington Post to pay for rent and food), put me to work on an unbiased guide to the issues around nuclear weapons, and introduced me to McGeorge Bundy (President Kennedy's national security advisor), Gerard Smith (who negotiated the first series of Strategic Arms Limitation Talks [SALT I]), Father Bryan Hehir (one of the authors of a famous pastoral letter from the U.S. National Conference of Catholic Bishops Letter on Nuclear Weapons), and many others. Within a few months, I had an offer to write a book for the publishing company owned by *Scientific American*. Elated, I set to work. It was 1982.

The first chapter laying out the problem almost wrote itself. But then problem after problem arose. The more I read, the less I understood. The months went by, and I struggled, trying more than a dozen different approaches but unable to bring anything into focus. For the second time, I had failed.

Eventually, I lost hope, gave up on the book, and got a job doing tech support at a company near Princeton, New Jersey. A few years later, I started my own small computer consulting company. I ended up doing work in three East Coast states, as well as California, and even Poland. But nuclear weapons wouldn't leave me alone. The problems, the mysteries nagged at me. At nights and on weekends, I read, made notes, and wrote. For decades. I got married and divorced. I lived at my parents' home for a time. The computer work dried up, and I worked for an attorney in Princeton. I got older. But steadily, late in the night and in the dim light of early morning, the work on nuclear weapons continued.

I read the classics of the field—Bernard Brodie, Thomas Schelling, Herman Kahn, and others. Sometimes several times. I bought books—

INTRODUCTION

secondhand whenever possible—and underlined, scribbled questions in the margins, and pondered. (I have nearly five hundred books just about nuclear weapons and related subjects now.) I wrote drafts of articles, even partial books, as I tried to work my way through various problems. Living my everyday life, I lived a second life separate from others where I wrestled (mostly alone) with nuclear weapons.

And slowly—painfully slowly—I began to see things more clearly. Quadrant by quadrant I drew my own map of the nuclear weapons issue, different from the map that most nuclear weapons policymakers had, different from the one taught at elite schools; but because I had been over the ground again and again, identifying the key features of the landscape, hacking pathways through seemingly impenetrable undergrowth, I was sure it was a true map. And in 2001, in a library in Nashville, I suddenly realized that the way to attack the problem—the approach I had failed to find all those years before—was through pragmatism. The central questions I was interested in were all about utility. I needed to use the intellectual tools that William James and Ludwig Wittgenstein had given me. Like a flash of light, I suddenly saw a way forward.

Then in 2007, I published a new and groundbreaking interpretation of the bombings of Hiroshima and Nagasaki. The article was accepted by Harvard's International Security, one of the two or three top security studies journals in the world. For a man with no master's degree, much less a Ph.D., it was a remarkable place to have your first scholarly article printed.

My second scholarly article, critiquing deterrence, won the McElvany Prize offered by The Nonproliferation Review for best essay on nuclear weapons and came with a $10,000 award and considerable recognition.

Then in 2010, I was given a large grant to write, travel, and speak about nuclear weapons. For six years I traveled the globe, eventually visiting twenty-two countries on five continents and meeting foreign ministers and diplomats, students and ordinary citizens, and everyone in between. I debated Sir Lawrence Freedman at Chatham House in London. I spoke at the Pentagon, the French National Assembly, the United Nations, the Scottish National Parliament, the U.S. State Department, Harvard, Stanford, Princeton, Georgetown, Yale, the Sorbonne, the U.S. Naval War College, King's College London, Hamburg University, Nagasaki University, University of Pretoria, the Mexican Foreign Ministry, the Belgian Parliament, the National Assembly in Costa Rica, and many other places.

IT IS POSSIBLE

And I wrote articles, reviews, and letters published in anti-nuclear journals like *Bulletin of the Atomic Scientists* and *The Nonproliferation Review*, in military journals like *Joint Force Quarterly* and *Revue de Défense Nationale*; and in foreign policy journals like *Survival* and *Foreign Policy*, as well as in the *Chicago Tribune*, the *New York Times*, the *Wall Street Journal*, and others.

In 2013, Houghton Mifflin Harcourt published my first book, *Five Myths about Nuclear Weapons*. I went on a book tour of cities in the United States. and spoke on radio and television. It was a time of heady success that I still think back on fondly. But eventually, the grants dried up and with them the ability to travel and speak. On the one hand, I wasn't radical enough for some anti-nuclear funders. They wanted moral arguments that focused on the horror of nuclear weapons. And on the other hand, I was shunned by the nuclear weapons policy-making establishment as far too radical—challenging some of their most cherished assumptions. For five years I struggled to pay the bills as I worked on a second book. And still I felt I hadn't untied the last tricky knot at the heart of the nuclear weapons issue. I sometimes felt that I would never figure the issue out. And, sometimes, in the darkest part of the night, I despaired.

*   *   *

Working on nuclear weapons for so many years has been a tempering experience. It has hardened me. And I'm still not sure how I was able to stick with it for so many years. What my mother told me that week in October 1962 changed me, I guess. But it didn't make me afraid. It didn't change my sunny outlook. It didn't break my will. It didn't make me into the guy carrying a sign that angrily declares, "The END is near!!!" What it made me, slowly, over the course of many years, was a realist.

That is not to say that I am a believer in the particular strain of academic thinking that some scholars call "Realism"—usually written with a capital "R." When I say "realism," I mean the ordinary, everyday version: an attitude that devalues fantasy, that prefers experience over sophisticated theory, that is willing to acknowledge the ugly and the flawed along with the good, and that takes facts with great seriousness. The kind of thinking that, applied with perseverance, often solves seemingly unsolvable problems.

I started out thinking about moral arguments, but I ended up prowling the corridors of a harsh and demanding realism. I am a man who uses realism to argue against nuclear weapons. And if that seems unlikely to you, believe me, it is not what I set out to do. But if you dedicate yourself to reality, it will teach you things. Not always what you expect, maybe not what you want, but occasionally what will serve you in good stead.

INTRODUCTION

So, that's why I am here. Now, what are you doing here? How is it that you find yourself reading a book on nuclear weapons when you probably live in a society that tries to draw a veil over the subject? I don't know you, obviously; but my guess is that you're here because you have a funny feeling, a kind of subliminal intuition, an unarticulated sense that there is something false about the life going on around us. You walk through everyday activities, but there is a kind of unreality to the world we've constructed. Occasionally you catch a whisper of something hard and important, but it's drowned out by the nattering of voices and seemingly endless entertainment. And you can't really get anyone to talk about it.

What brought you to this book is that you know something. You may not have put words to it, but you sense it. You can feel the chasm between the dangers that are out there and the apparently untroubled way we live. It's as if we are sleeping (or perhaps dreaming) in some sort of comfortable cocoon—a cocoon that prevents us from feeling the things that are wrong but that won't protect us if anything dangerous happens.

* * *

This isn't normal. People don't normally ignore important, life-threatening issues. How did we get here? How did the beliefs that most people accept about nuclear weapons come about? Here's what happened. In the late 1940s and early 1950s, a small group of experts and government officials in the United States was asked to come up with policy recommendations about nuclear weapons. They had to work with what they had, which wasn't much. They were told that the weapons were the most powerful ever invented, so powerful they would revolutionize everything. And to go with these claims, there was only a single piece of real experience, one solid data point: The weapons had been used twice and seemed to have won a war overnight. So, these experts and government officials made assumptions and came up with policy suggestions. But (probably unbeknownst to them) their thinking was distorted by powerful feelings of fear and awe. They were, after all, working in the first years of the Cold War. What they concluded was that nuclear weapons are necessary for the survival of the United States, and they are likely to be around for as long as there is danger in the world. In other words, they estimated the value of nuclear weapons as very, very high.

Over time, to justify that estimate, they developed abstract theories, used the new intellectual technique of game theory to think about future wars, developed so much jargon that it almost became a language of its own, and insisted that the subject was so important and so complex that only those with Ph.Ds. from elite universities could participate in the conversation. And because they had been asked to do a highly uncertain thing where the stakes were impossibly high, they eventually developed the emotional protection of insisting that they were certainly right.

And ordinary people had no way to judge what the experts and government officials said. "Countervailing strategy," "ALCMs," "B61 mod12," "compellence"—it was incomprehensible. How were they to know? So, ordinary people said to themselves, "Well, nuclear weapons are enormously big, and everyone says they're the ultimate weapon. So, that kind of makes sense." And they said, "They're the most destructive weapons ever built by humans, and there's no arguing that destruction plays a part in war. I guess that lines up." And they had all heard of Hiroshima—they knew that an earlier version of these weapons (atomic bombs) had been so effective they forced Japan to surrender in just four days at the end of World War II. And they had to admit, "When you stop and think about it, there haven't been any major wars over the last seventy years—not World War II–scale wars, anyway. So maybe this deterrence stuff does work like they say." And in the end, they thought, "Hey, I'm no rocket scientist—who am I to judge?"

And again and again, the experts said that nuclear weapons would always be with us, which meant there wasn't anything anyone could do about the danger. So even if they were worried, ordinary people tried not to think about it. Can you blame them?

The problem is that the experts' first conclusions, the foundations that all the subsequent thinking was built on, were just an estimate; and those first conclusions haven't been reevaluated or reassessed since. The nuclear weapons policy elite work in isolation, and over time the people in it have come to think of themselves as the only people who are smart enough and tough enough to think clearly about this issue. In their view, any thought of eliminating nuclear weapons is merely wishful thinking, a kind of foolish idealism. As a result, they stopped listening, they talked only to each other, and that constant self-reinforcement locked them deeper into their own peculiar view of the world.

* * *

Imagine what it would be like to live in a world without nuclear weapons. It would still be a world with risks and with war. But it wouldn't be a world where adults have to tell their children that everything might someday be gone. It wouldn't be a world where catastrophe seems to loom so large in our collective imagination, where every other video game or blockbuster movie seems to be set in a world that has been devastated and destroyed.

INTRODUCTION

The threat of nuclear war sits on the horizon like a huge storm, darkening the sky—clouds banking up higher and higher, winds rising, with a continual distant grumble and shudder of thunder. Having that looming threat always on the horizon affects how we feel. Imagine a blue sky from horizon to horizon, not a cloud to be seen. Remember the way those clear, sunny days make you feel. That's what a world without nuclear weapons would be like.

"But how would it be possible to get rid of nuclear weapons?" someone might ask. "It wouldn't be safe to get rid of them." A lot of people have somehow, for reasons I don't really understand, come to believe that nuclear weapons can only be eliminated if the United States lays down its weapons. This act of altruism (so this line of thinking goes) will generate a feeling of trust among our adversaries, and they will then magically become good and kindhearted human beings who will gladly lay down their nuclear weapons as well. This obvious nonsense is a ridiculous way to think about eliminating nuclear weapons. If that's the sort of book you're looking for, go someplace else. Hoping against hope that human beings can change their essential nature is an approach that I believe is bound to fail.

But that's not how nuclear weapons will go away. They'll be eliminated by a treaty that forces everyone to get rid of their nuclear weapons at the same time. There's no other way. "But how could you get everyone in the world to give up nuclear weapons?" someone might ask. First, you don't have to get everyone to give up the weapons. You only have to get the nine countries that have them to give them up. And the key is not what's in the treaty that eliminates the weapons. The key is what happens before the treaty. The key to eliminating nuclear weapons is to get people to see them differently, to imagine them differently, to think about them differently.

Right now, the biggest obstacle to eliminating nuclear weapons is how much people value them. It is the value that we attach to them in our heads that is the barrier. Think about it. If you were standing on your front step with a priceless diamond in your hand—beautiful, sparkling, very, very valuable—and I came over to you with a diamond in my hand and said, "Hey, couple of other folks I know who own incredible, big diamonds are going to drive over to the dump with me and throw our diamonds away. You wanna come?" You'd look at me like I was crazy.

But imagine that you looked down at the diamond in your hand and suddenly—as if a mist had cleared from your eyes—you realized that it wasn't a diamond at all. It was just a lump of coal. You look down and say, "Whoa. This isn't a diamond. It's a lump of coal!" (We often say the obvious when we're surprised.) And then imagine that I came over to you and—showing you the lump of coal in my hand—said, "Hey, some of the neighbors and I are going to go over to the dump and throw these lumps of coal away." You'd say, "Let me get my keys."

People believe—down to their toes—that nuclear weapons are priceless, peerless weapons that have a power approaching that of gods. Nuclear weapons are, in their heads, like diamonds. But what if those diamonds turned out to be fake? What if someone had crept in at night and taped little realistic-looking pictures of diamonds over lumps of coal? Wouldn't that change everything? What if when people thought about nuclear weapons, the image that leaped to mind was a lump of coal instead of a diamond?

Getting rid of something valuable is very hard. Getting rid of something undesirable is easy. What if nuclear weapons, instead of being something valuable, were something undesirable? I'm not talking about their dangers here. I'm talking about their utility. What if nuclear weapons weren't that useful? If nuclear weapons are valuable, then attempts to reduce their numbers (or even get rid of them) are like trying to pry people's fingers loose from a priceless diamond. This is the approach we've been using for seventy-five years. It hasn't gone so well. But if you could convince people that nuclear weapons aren't that useful and are also very, very dangerous, they would relax their grip of their own accord. This is what I'm proposing.

Think about how hard it's been to negotiate even modest reductions in the number of nuclear weapons—long, drawn-out talks, often taking years at a time, haggling over technical details, experts arguing and arguing. And even when a treaty was finally arrived at, sometimes it wasn't ratified at all because some people objected that the treaty was giving up too much. This is what it's like to try to pry people's fingers apart.

Compare that with the treaties to ban chemical and biological weapons. Those two treaties only took a short time to negotiate; they were approved without a murmur. Why? Because World War I demonstrated conclusively that chemical weapons spread by the winds just weren't that useful. They were hard to aim and control, they had the potential to kill your own soldiers, and they didn't deliver a strategic advantage to either side. It was clear that, although they were dangerous, chemical weapons weren't that valuable as weapons. And the COVID pandemic has shown how hard it is to control biological weapons. A virus escapes from a lab or jumps from animals to humans, and it is almost impossible to stop its spread. Three years later, eleven million people are dead worldwide. Biological weapons would be hard to control, even harder than chemical weapons.

These realities made the negotiations to ban chemical and biological weapons go relatively swiftly. The key was that people didn't value chemical or biological weapons; they had realized they were ugly little lumps of coal,

INTRODUCTION

not diamonds. To eliminate nuclear weapons, therefore, the first and most important step is to show that nuclear weapons have been overvalued.

*   *   *

I understand that most of you, right now, don't believe that it is possible to transform nuclear weapons from diamonds into coal. But it is possible. This book explains how.

The first thing we're going to do is address the more than seventy years of failure that anti-nuclear protestors have had. This failure makes some people ask, "If you can't get something done in seventy years of trying, isn't that pretty good evidence that it's impossible?" And they have a legitimate point. Long-term, repeated failure is often an indicator that what you're trying to do can't (realistically) be done. But it turns out the long record of political failure for the anti-nuclear movement is not an indicator of the truth or falsity of their arguments. In fact, it doesn't tell us anything about the value of keeping or discarding nuclear weapons, because the game was rigged. One side in the policy debate stacked the deck. Early on, proponents of nuclear weapons framed the issue in a way that ensured they would always win. So, in this first essay, we'll unpack how the politics of the debate got rigged and how to unrig it.

The second essay addresses another source of skepticism: the society-wide acceptance of nuclear weapons. Presidents, bureaucrats, admirals, journalists, pastors—the broad majority of people in the United States have accepted the necessity of keeping nuclear weapons. And then beyond the United States in the United Kingdom and Russia and India and many other countries as well, people have accepted the necessity of keeping nuclear weapons. And they've held these views for decades. If so many people, from so many countries, in so many different walks of life, for so many years have accepted that nuclear weapons are here to stay, how could they be wrong? Could a society-wide consensus, shared by people in many different countries, be wrong? This second essay shows, using historical evidence, that such a society-wide consensus is no guarantee at all. In fact, there is at least one similar example of this kind of shared belief being dangerously wrong.

In the third essay, we'll talk about what happened. Those first experts and government officials overestimated the value of nuclear weapons, and that overestimation still lies at the heart of all our troubles. To straighten out the present, we must understand the past.

The fourth essay lays out the plan of attack for eliminating nuclear weapons. It starts, as you might expect, by showing that the question of elimination has been framed in such a way that it only has one answer: It's impossible. But if we step back and think about the way all technology comes and goes, if we challenge the framing and substitute a much more realistic approach, the problem changes dramatically.

Fifth, with the preliminaries over at last, we'll turn to the characteristics of the weapons themselves. And we start with the first intuition that affects attitudes toward nuclear weapons—an intuition that needs to be unearthed, examined, and its falsity laid bare. Most people intuitively believe that because nuclear weapons are the biggest weapons ever built, because they are the most destructive things on the face of the earth, they are necessarily the best, most effective, most decisive weapons of war, too. We'll talk about why the biggest weapon isn't always the best one.

Sixth, we'll discredit the crucial first impression of nuclear weapons—the first, shocking success of nuclear weapons, the event that set the tone for everything that followed: the sudden surrender of Japan after two of their cities were attacked with nuclear weapons. I'll present a powerful rebuttal to the argument that the first time these weapons were used, Japan's leaders were so shocked they surrendered in just four days.

Seventh, we'll go to the heart of the matter and discuss the overwhelming evidence that nuclear weapons not only are poor weapons, they are hardly useful for any military task at all. Far from being the ultimate weapon, nuclear weapons are almost entirely useless—not because of moral restraints but because, even viewed coldly and entirely pragmatically, there are hardly any situations in which they help you win.

Eighth, we'll examine the claim made by some people that even though nuclear weapons aren't very efficient weapons on the battlefield, their very destructiveness is what makes them important in war. Nuclear weapons are effective, so this argument goes, because they are overly destructive, because they kill civilians, because, in other words, they are cruel rather than efficient. To strengthen the case for elimination, we will discredit the argument that cruelty is an effective way to win a war.

In the ninth essay, we'll try to figure out how to explain why, if they are not very good weapons, nuclear weapons seem to be so important. Why do they dominate the world stage in the way they do? It turns out there is an obvious way for nuclear weapons to play a large role in the life of societies without being great weapons.

In the tenth essay we'll talk about what most nuclear weapons advocates claim is the most important subject: nuclear deterrence. Nuclear weapons, so the claim goes, keep us safe, solidify alliances, and make us influential. And what makes all these benefits possible is nuclear deterrence—the

INTRODUCTION

threat of retaliating with those weapons if we're attacked. If nuclear deterrence is so useful and powerful, why would we ever want to get rid of nuclear weapons? This essay examines nuclear weapons deterrence theory and shows that it is inconsistent, illogical, and out of step with the latest knowledge about psychology, neuroscience, and human nature.

In the eleventh essay we examine the history of nuclear deterrence, the practical reality on the ground, and show that nuclear deterrence is inherently flawed and will inevitably fail.

Finally, we'll talk about what steps need to be followed to turn the information and arguments in this book into actual elimination.

* * *

The war in Ukraine has highlighted the role that nuclear weapons play in our lives. Russian leaders have issued a series of nuclear threats as the war has gone on.

Four days after Russia invaded Ukraine on February 24, 2022, President Putin ordered Russia's nuclear forces onto high alert. On April 24, 2022, Russian foreign minister Sergey Lavrov said that the support that other countries were providing to Ukraine could lead to a world war in which Russia would use all of its arsenal of weapons. Three days later, Putin, addressing Russia's legislature, said that Russia would respond to any military provocation from outside Ukraine and that it might do so with nuclear weapons. On September 27, 2022, deputy chairman of Russia's Security Council, Dmitry Medvedev, said that Russia had the right to defend itself with nuclear weapons if it is pushed beyond its limits and that this "is certainly not a bluff." On November 5, 2022, Putin, during a meeting with French President Emmanuel Macron, referenced the bombings of Hiroshima and Nagasaki and said they demonstrated that "you don't need to attack the major cities in order to win."

This series of threats, which must be taken seriously, has shaken many people in the United States and elsewhere. The comfortable myth that the Cold War was gone and that nuclear weapons no longer posed any real danger—that they were on a long, slow path to elimination—evaporated.

What the war in Ukraine demonstrates is that as long as nuclear weapons exist, as long as nations have grievances, and as long as nations have unfulfilled ambitions, the danger of nuclear war will also exist. For two decades after the end of the Cold War, we pushed the issue out of our minds. Of course, the danger was there all along. But for a while, we lived in comfortable denial. The war in Ukraine has made the issue of nuclear

IT IS POSSIBLE

weapons unavoidable. This is a problem we cannot ignore. It is a danger we cannot wish away. You are living with this every day of your life right now. Nuclear weapons pose an ongoing and deadly threat. They must be dealt with.

I am going to ask you to face some difficult facts, some of which you probably won't like. I'm going to challenge some ideas about the way the world is that you've gotten accustomed to. There are harsh realities that need to be acknowledged, realities like "human beings are capable of appalling acts of cruelty"; "we are not much more advanced, morally, than the human beings who lived thousands of years ago"; and "despite the scar that the destruction of a city leaves on a society, destroying cities doesn't win wars." It won't necessarily be easy. Nuclear weapons have wrapped us in a cocoon of self-regard, of false security, and of complacency. Eliminating nuclear weapons will require crawling out of those warm, unrealistic cocoons and leaving them behind.

And I am going to ask you as well to do something that many people find emotionally difficult these days. I am going to ask you to hope. After years of just skating by and ignoring the danger, after decades of passively accepting that nuclear weapons are here to stay, we've grown used to living under a menacing cloud, with no hope that the danger could ever lift. I'm going to ask you to believe that nuclear weapons could actually go. Some folks will find it hard to believe that, to imagine a future so different from the one we are living in. But if we want to eliminate nuclear weapons, we must believe. Nuclear weapons cannot be eliminated without hope.

But there's also good news. If we face that reality, and if we do the work necessary to get others to face it, it will be relatively easy to get rid of nuclear weapons. Once you change the image in people's heads, they won't just be willing to get rid of nuclear weapons; they'll be eager to. Who wants to keep an ugly, dirty piece of coal? Undermine the importance of nuclear weapons, reduce their perceived value, and elimination gets much, much easier. It's not that hard to get someone to toss away something they think is worthless.

And then we can put the fear that nuclear war is inevitable, the fear which has undermined our courage and sapped our hopes, behind us. We can live clean, breathe deeply at last, feel the sunshine, make things happen, and perhaps even discover paths that no one has yet dreamed of.

It is possible.

INTRODUCTION

1. I am aware of the irony involved in having just said in the previous paragraph that I don't think stirring emotion is the way to solve the problem, and in this paragraph, I'm relating how descriptions of the horror stirred me to action. Life isn't simple.
2. Constantine I converted to Christianity in 312 CE.

1. THE GAME WAS RIGGED

It is possible because the long record of failure doesn't count—the game was rigged.

The record of failure by those who oppose nuclear weapons is more than seven decades long. The failure of anti-nuclear think tanks, nongovernmental organizations, and activists that have worked against nuclear weapons is undeniable. Almost from the moment nuclear weapons were first used, people have been protesting against them, raising doubts about the wisdom of keeping them, and decrying their horrible destructive power. But those efforts have barely dented the belief that there will always be some nuclear weapons. Perhaps they will be small arsenals, but there will be arsenals. The anti-nuclear movement's best successes, in terms of popular organizing, came in the early 1980s, when anti-nuclear referenda were passed in about a third of the United States and more than a million marched and protested in Europe. But that movement in the 1980s, for all its success at pressuring the European governments and the Reagan administration (and even though it attracted millions of people), was not strong enough to advocate the complete elimination of nuclear weapons. The underlying rationale for keeping nuclear weapons has never been seriously threatened.

It is true that the arsenals had swelled to monstrous proportions in the 1970s and 1980s. It is true that they have steeply declined since. The seventy thousand weapons that were in existence in the middle 1980s have shrunk down to something like thirteen thousand weapons today. That is real progress, and almost certainly some of that progress is due to the efforts of groups and organizations in the anti-nuclear community.

And today, the efforts of the International Campaign to Abolish Nuclear Weapons (ICAN), the Red Cross, a collection of dedicated diplomats, and other groups led to a treaty that commits each country that signs the treaty to ban nuclear weapons. *The Treaty on the Prohibition of Nuclear Weapons* (TPNW) has had a worldwide impact and changed the conversation about nuclear weapons. Before this effort, most countries that didn't have nuclear weapons were excluded from discussions about these weapons. The nuclear-armed states condescendingly told the rest of the world to leave this problem to them. A nuclear war would have worldwide consequences, but apparently, that was no reason to allow non–nuclear-armed states to have a say in the matter. Since the TPNW was passed by the United Nations in 2017, however, the non–nuclear-armed states have become increasingly vocal in their objections. There is now real international pressure on nuclear-armed states to abandon nuclear weapons. Every country that signs on to the treaty (and there are now more than eighty countries that have signed) increases that pressure.

But set against this progress is the fact that the nuclear-armed states are more committed to nuclear weapons than ever. The long, slow decline of nuclear forces that took place from the 1980s to the 2010s has now been reversed. Each of the nine nuclear-armed states is now either expanding its arsenal or upgrading its existing weapons. We have moved from a period of fading interest in nuclear weapons to a second nuclear arms race. Although there are only thirteen thousand nuclear weapons in the world today, that number is increasing and will likely continue to grow. And the number of countries that possess nuclear weapons appears likely to expand as well.

The real battle, the one that will determine whether nuclear weapons continue to exist in our world, the one that the anti-nuclear groups have not figured out how to win, is the battle over the rationale for keeping nuclear weapons. Should nations continue to rely on nuclear weapons? That is the central question. But what this second arms race shows is that that rationale is still roughly as strong and as effective as it was when it was first articulated in the Cold War, nearly seventy years ago.

FRAMING

Polls show that most Americans would get rid of nuclear weapons (if they felt they could). But for the most part, people try to put nuclear weapons out of their minds. They don't just ignore the issue; they actively try not to think about it. The unwillingness to think about the issue, however, is unsurprising after all these decades of failure.

The only way this record of failure could be explained away, the only way you could potentially break through people's learned indifference, would be to claim that there is some extraordinary, previously unnoticed reason for this long string of defeats. To open people's minds to the possibility of elimination, you'd have to somehow show that there has been some sort of sub-rosa chicanery going on, some form of cheating hidden in the debate that has been tilting it—putting a finger on the scale, so to speak—so that the advocates of nuclear weapons keep winning. Without this kind of unlikely explanation, a realist would have to admit that the chances of eliminating nuclear weapons are almost nil.

But it turns out there is just this sort of hidden tilting of the playing field. Buried in the debate, in a way that you have probably never noticed, there is a built-in advantage for the people who argue that we should keep nuclear weapons. It has to do with framing—political framing: the art of setting up a debate in such a way that the answer is determined by the way the question is phrased, defining the terms of a debate in such a way that only one side can possibly win.

A friend of mine once talked, with a shake of his head, about the brilliance of the anti-abortion movement labeling themselves "pro-life." "Who could be against life?" he asked sadly. By framing the question of whether women should be allowed to have a medical procedure that ends a pregnancy as a choice between being in favor life or being in favor of killing, people who were against abortion gave themselves an enormous advantage.

An older and more obvious example of framing is asking, "Do you still beat your wife?" The question forces you to either admit that you are still beating her or say that you used to beat her but now you've stopped. Either way, if you respond to the question, you're trapped.

The nuclear weapons debate got framed in this same way at the very outset, and the impact of that clever maneuver is still being felt today. To see how the advocates of nuclear weapons got an unbeatable advantage, let's revisit some long-forgotten history.

THE FIRST TWO DEFEATS

Almost immediately after reports reached the United States about the destruction of Hiroshima and Nagasaki, Americans of all kinds—conservatives and liberals, young and old, hopeful and cynical, and in every kind of occupation—began worrying about the long-term consequences of the existence of such devastating weapons.[3] It seemed that these weapons could one day destroy the world, and groups sprang up determined to head off that destruction by eliminating the weapons entirely. And each group had a different plan. The first two plans that gained national attention, however, turned out to be the most important ones. Together, they had a remarkable, long-lasting impact. That impact, though, wasn't exactly what the authors of the proposals hoped for.

SHARING THE SECRET

The first plan was supported by scientists and called for resolving the danger by copying the approach used in science. In the world of science, when you make an important discovery, you tell everyone about it. This has three results. First, you get credit for the discovery—if you publish first, then there's no doubt whose work it is. Second, everyone in the field can then test your results and try to reproduce them. This serves as an important double-check to make sure your results are correct. And third, everyone else can then take what you've learned, add it to what they know, and push the boundaries of discovery even further. Publicly releasing your findings allows everyone to, in effect, work together to push knowledge forward as quickly and efficiently as possible.

To see this kind of information sharing at work, you need only look at early efforts to split atoms. Progress came from many countries, and from scientists of many nationalities. An Englishman, James Chadwick, discovered the neutron in 1932. In 1933, Leo Szilard, a Hungarian, conceived the notion of splitting atoms with neutrons to produce a chain reaction. In 1938, Otto Hahn and Fritz Strassmann, German chemists, bombarded a solution of uranium nitrate, hoping to create heavier new elements but instead found much lighter barium in their solution. They wrote a former colleague physicist Lise Meitner, an Austrian Jew who had fled Austria for Sweden. With her physicist nephew Otto Frisch, Meitner worked out that to create barium, the uranium atoms must have actually split apart. And so on. Scientific understanding was an international product, with many contributors. Progress was rapid, and the benefits were available to all.

The first proposal for getting rid of nuclear weapons argued that rather than keeping the science behind nuclear weapons a secret, the information should be shared with Russian scientists. This act of scientific openhandedness, according to the plan's proponents, would build confidence between scientists in the United States (and its allies) on the one hand and Russian scientists on the other. And that confidence would lead to greater trust between the two nations. Instead of competing with one another to build the biggest nuclear weapons—in other words, instead of starting an arms race—the United States and Russia could cooperate, learn to trust one another, and thus begin the process of bringing peace to the post-war world. Scientists would lead the way. This suggestion eventually made its way to the United Nations as a major disarmament proposal that was called the Baruch Plan.

ONE WORLD OR NONE

The second proposal for eliminating the danger of nuclear war was a call to rapidly institute a world government. Only a strong centralized government could prevent the kind of competition between nations that led to war, so the reasoning went, and therefore it was imperative to put an end to such dangerous competition. Only by empowering the United Nations, or some sturdier successor, to judge and settle disputes without war could the world be saved from nuclear war. One of the strongest formulations of this proposal was the book, *One World or None*, whose title cleverly encapsulated the argument that if world government were not instituted quickly, the world would be destroyed by nuclear war.[4]

These two proposals each received widespread attention, were debated in newspapers and magazines and became a part of the national conversation about nuclear weapons. They also helped to fix the shape of the debate for decades to come.[5] Even though they were widely discussed, and even though both garnered considerable support in various quarters, neither ever amounted to anything. The problem was that both were hopelessly naive.

Superiority in military technology has often been the basis for military dominance. In the ninth century BCE, the Assyrians, for example, conquered much of the present-day Middle East by taking advantage of iron weapons, at the time a new technology. Their opponents, who were armed with softer bronze weapons, could not match the Assyrians' newer, stronger, cheaper weapons. In the eleventh century, the Mongols built the largest empire the world has ever known by relying on advances in bow technology. The ability to shoot farther and with greater force (combined with superior horsemanship, unbending discipline, careful intelligence gathering, and ruthless leadership) was the basis of one of the greatest feats of military conquest in recorded history.

The problem with sharing the secret of the bomb was that the science behind nuclear weapons was not an esoteric branch of human understanding. It made possible military tools that seemed to promise the ability to conquer on a terrifying scale. No sensible statesman shares important military technology with an adversary. Military advantage could be the difference between national survival and national destruction. And, in fact, we now know the plan was doomed from the start. New information revealed when Soviet archives were opened in the 1990s shows that Joseph Stalin, the Soviets' leader "was ready to thwart the Baruch Plan long before it was announced."[6] Given Stalin's suspicious nature, he seems to have judged that a naive peace plan couldn't possibly be serious and therefore must be the mask for a more sinister move. The notion of sharing the secret of the bomb never had a chance. Sharing what might be dominant military technology in order to build trust was a well-intentioned but foolish idea.[7]

The proposal to build a world government was also flawed. A successful government of laws, it has been persuasively argued, must be based on shared values.[8] Laws are merely words—powerless in themselves—unless they reflect underlying beliefs that have taken root throughout a society. As the attempt to prohibit alcohol in the United States in the 1920s so vividly demonstrated, it does no good to pass even a constitutional amendment if a large and determined segment of the population does not believe in the underlying goal of the law. Prohibition led to widespread flouting of the law by ordinary citizens, the formation of gangs competing to supply liquor, and eventually widespread lawlessness that nearly undermined the rule of law itself.

If shared values are essential for governing a country, then in order to build a world government, it would be necessary to discover values that are shared worldwide. Yet the world is a jumble of differing cultures, religions, business practices, laws, customs, languages, ideas, and attitudes toward life. Finding commonly held values that would be sufficient to construct the rule of law over the entire world seems like an extraordinarily complex and demanding task. (It might, of course, be possible to impose a set of values on the world by force—for a time. But eventually the human desire to live one's life as a reflection of one's own inner values would lead to insurrection, violence, and revolt.) Of course, a world government may one day be possible, as powerful communication technology brings people closer together and a true world culture grows; but it is surely far beyond our reach today and for the foreseeable future. Imagining that world government could have been instituted in the late 1940s was a utopian dream.

So, each of the first two suggestions for eliminating nuclear weapons were, in different ways, flawed from the start. This is not particularly surprising. When you're faced with a complex new problem, it often takes a few tries to come up with a workable solution. But these first two failures had an unexpected and outsized impact on the debate.

In those early days of thinking about nuclear weapons, there were no thumbnail sketches of the "typical" nuclear weapons opponent or advocate. The two sides hadn't hardened into set positions. But all that changed after these two ill-considered proposals.

The advocates of nuclear weapons didn't see these first proposals as simply a pair of ideas that didn't work out. They didn't see them as efforts to find a solution to a very difficult problem that (unsurprisingly) fell short. Advocates of nuclear weapons looked at the two proposals and saw a pattern, a pattern that they then used to recast the entire debate.

They pointed out that the two proposals shared a common characteristic: they were both solutions that relied, to an unusual degree, on idealism. Does the example of good behavior inspire good behavior in others? In an ideal world, it does. Is it possible for very different peoples to share common values? In an ideal world, it is. Advocates of nuclear weapons took that similarity and made it into an iron rule. They claimed that these two proposals proved that there was only one way to think about eliminating nuclear weapons. And more than that, there was only one way to think about the people who said we should eliminate nuclear weapons. Since both proposals were naive and idealistic, they concluded, it must be the case that any proposal to eliminate nuclear weapons was necessarily idealistic. Opposition to nuclear weapons was informed by a kind of simple-minded, utopian reasoning, they decided. Anyone who opposed nuclear weapons was, therefore, an idealist. They were all soft-hearted, addle-headed peaceniks. It was regrettable that so many people shared these views, and although they shouldn't be punished for holding them, the country had to be protected from these fools, naïfs, and hopeless Don Quixotes. Former U.S. Secretary of Defense James Schlesinger's dismissal of opponents of nuclear weapons, for example, is typical. "The notion that we can abolish nuclear weapons reflects a combination of American utopianism and American parochialism.... It's like the *Kellogg-Briand Pact* renouncing war as an instrument of national policy.... It's not based upon an understanding of reality."[9]

It didn't take long for the belief that all opposition to nuclear weapons was idealistic to harden into an established "fact." And this view has dominated the thinking of policymakers for more than seventy years.

Once the opponents of nuclear weapons had been labeled as idealists, it suddenly became clear to advocates for nuclear weapons who they themselves were. The identity of proponents of nuclear weapons was created, in part, in contrast to who they were not: They were not the people who wanted world government or to give away the secret of the bomb. If the people who wanted to get rid of nuclear weapons were hopeless idealists, then the people who wanted to keep nuclear weapons must be realists. They must be hard-hearted, tough-minded people who were unafraid to see the world the way it really was. They had to be people who understood that to survive in a world of nuclear weapons, you had to be tough.

That toughness was perhaps best exemplified by General Curtis LeMay, the Air Force general who had commanded and organized the U.S. campaign to bomb Japanese cities in World War II and who went on to be Chief of Staff of the Air Force. LeMay argued that nuclear weapons were decisive as long as you weren't squeamish. And nuclear weapons advocates agreed. Only the people who were tough enough to embrace such harsh positions, they decided, were capable of being realistic about the issue.

REALISTS VERSUS IDEALISTS

Once the roles of the two different sides had been sorted out, the advocates for nuclear weapons believed it was possible to see the shape of the entire debate. Based on only two pieces of evidence (and perhaps a bit of wishful thinking), they leaped to the conclusion that the struggle to make nuclear weapons policy was a contest that pitted realists against idealists. The realists saw the world as it actually was; the idealists saw the world the way they hoped it would be. And once you understood this, it became crystal clear who was right about nuclear weapons. The realists were right. And they would always be right. It can be inspiring to think about how the world might be, but foolish dreams can get you killed. To survive, you had to see the world the way it really is. You had to hold on hard to realism.

You can see immediately why this had such a huge impact. In fact, it assured that the advocates of nuclear weapons would win every debate. When there is a danger of losing your life, which would you rather rely on an idealistic plan or a realistic one? When heavily armed bandits surround your party on a lonely mountaintop, are you more inclined to follow the lead of the person who says, "We should depend on the inherent goodness of human beings and invite them in, show them our trust, and then they will surely return that trust by letting us go"? Or the one who says, "I think there's a path off this mountaintop that only a few people know about. If we can stealthily kill any bandit guarding it, we can make our escape"?

Where there is danger and the possibility of death on the line, most people want a realistic plan of action.

As a result, once the debate was defined as a contest between realists and idealists and the people who said we should eliminate nuclear weapons were branded as the "idealists," their proposals for dealing with nuclear weapons were largely dismissed out of hand—not just by nuclear weapons advocates, but by most ordinary people, too. Everyone can see the danger that nuclear weapons pose. Who wants to risk trying an idealistic plan when faced with danger of that kind? Within a few years, this framing of the debate became the standard view in American society. There were "realists," who said (reluctantly) we have to keep nuclear weapons, and "idealists," who had foolish utopian ideas about getting rid of them. After becoming a society-wide consensus in the United States, it became a consensus in the nuclear-armed states worldwide. In some ways, it is remarkable that proposals for reducing nuclear weapons have done as well as they have given that almost everyone automatically assumes that eliminating nuclear weapons is a ridiculous, unrealistic proposal.

The "realists" who advocate for nuclear weapons have an enviable record of success in policymaking. They almost always get their way. And this experience has played a part in their developing a calm sense of their own rightness. They seem to imagine that their position is unassailable. When they speak, they sometimes sound like frightening old judges: official, remorseless, beyond doubt.

They seem formidable, but they have a weakness. Despite their many victories, there is a crucial frailty in their thinking. True, they were right to oppose the idea of sharing the secret with the Russians. It was a foolish proposal. And the notion that human beings are ready for full world governance was also doubtful. But those two early attempts are not the only way to go about elimination.

The people who advocate keeping nuclear weapons are actually quite human. They have strong emotions, and those emotions have led them to take on flawed position after flawed position and treat each one as cold, hard fact. They think they are objective, but their self-evaluation is myopic. They are fearful rather than analytical. They exaggerate and distort rather than seeing the world clear-eyed and objectively. The fatal flaw in their positions, ironically, is that they are not realists.

Nuclear weapons advocates are so sure of their realism that they don't bother to go back over their reasoning and reexamine their assumptions. But that sort of confidence is often the first indication of a problem. After so many years of success, they act today almost like people in the grip of a divinely inspired faith. They are not angry with the opponents of nuclear weapons but saddened and gently dismissive. They see the opponents of nuclear weapons as well-meaning but wrong-headed, rather like gentle, long-haired, patchouli-scented, pot-smoking hippies—for the most part harmless, but they shouldn't be allowed to interfere in the discussions or decisions of serious, responsible people.

Nuclear weapons "realists" don't understand that they have been swept up by deep emotions. They don't see that their patina of realism covers assumptions and ideas that are, at best, wrong and, at worst, dangerous. Their entire position is doubtful, but they don't see it.

NO MORE WAR

This is rather strong criticism. The nuclear weapons advocates' position is so dominant today that challenges to its fundamental assumptions hardly ever occur. It might be hard for some people to imagine that people who advocate for keeping nuclear weapons can be unrealistic. So, let's detour for a few minutes to examine one case where their failure to adhere to realism is clearly on display.

Nuclear weapons advocates assert that nuclear weapons prevent major wars. And they can bring some evidence to bear in support of this claim. For more than seventy years, there has been no war between the United States and Russia—or between any other major nuclear powers. Nuclear weapons advocates extrapolate from this evidence that nuclear weapons are instruments of peace and therefore profoundly useful. John Lewis Gaddis, the foremost historian of the Cold War, for example, sums up this conclusion succinctly: "As the means of fighting great wars became exponentially more devastating, the likelihood of such wars diminished, and ultimately disappeared altogether."[10] Gaddis is saying that "great" wars have disappeared altogether and nuclear weapons are the reason. This claim is one of the linchpins of the nuclear weapons advocates' position. After all, if major wars can still occur, then the possibility that nuclear weapons will get used is also greater, and that makes it dangerous to keep them. This question of whether major wars between nuclear-armed opponents are possible is one of the most important in the discussion.

But the nuclear advocates' position cannot be right. Examine the history of our civilization—which Winston Churchill once called "the dark lamentable catalog of human crime"—and you'll find that we have been fighting wars with dogged persistence for at least six thousand years. Or as President John F. Kennedy put it, "[T]he human race's history, unfortunately, has been a good deal more war than peace."[11] Every era of history has experienced war with disheartening regularity. There are sometimes pauses and respites, but the lust for war always reemerges.

Is this desire for war instinctual? Is it deeply imprinted on our brains and hearts—an inescapable imperative? Or is it merely an attitude that is learned—a cultural phenomenon that results from the competition for scarce resources? Experts disagree about how exactly the lust for war holds us in its grasp. And they disagree over how strongly it grips us. Those who have lived through long periods of peace tend to be optimistic. They believe the hold of war can ultimately be overcome. So, for example, Norman Angell, a British M.P., wrote a best-selling book in 1910 that optimistically argued (four years before a war that convulsed Europe and slaughtered millions) that war could be rendered obsolete.[12]

Those who have lived through major wars tend to take a darker view. American philosopher William James (whose brother Garth was severely wounded in the American Civil War) once wrote, "Our ancestors have bred pugnacity into our bone and marrow, and thousands of years of peace won't breed it out of us."[13] As a student of history, I side with James. Over the very long run, it may be possible to finally abolish war; but if humans were to suddenly give up fighting wars, it would be a monumental change—a revolution in human behavior. Losing our taste for war would be like renouncing our predisposition for religion, our love of beauty, or our tendency to overeat. It would mean that a trait that had been an integral part of human character for thousands of years had somehow been forsaken. The evidence of history makes plain that humans fight wars regularly and that those wars often become savage, all-out affairs. When they say that nuclear weapons have caused human beings to cease fighting all-out wars, what the advocates of nuclear weapons are claiming is that nuclear weapons have somehow, mysteriously, irrevocably altered human nature. Nuclear weapons, according to nuclear "realists," have somehow permanently suppressed the heretofore unquenchable desire for all-out war.

It's a remarkable claim, a surprising claim, one that runs counter to a long-accepted vein of scholarship. Sigmund Freud, for example, wrote in an essay called "Why War?" that "there is no likelihood of our being able to suppress humanity's aggressive tendencies."[14] Imagine if someone walked up to you on the street and said, "I've invented a new technology that will permanently rid people of their desire for religion." You'd immediately be skeptical, wouldn't you? Or perhaps, given the current tendency to believe that technology can do anything, you wouldn't be. But very few pieces of technology have ever created a change this radical. Even the slickest, newest smartphone has yet to change the nature of love, the closeness of family ties, or our tendency to coalesce into tribes. It is a long-standing truth that technology rarely changes human nature.

Perhaps, over the long arc of history, you could argue that some pieces of technology have shaped who we are. The wheel, for example, made travel and movement so much easier that it may well have changed how we understand our own rootedness to place or home. And fire, certainly, has shaped the nature of human civilization by making it possible to turn otherwise tough or diseased plants and animals into edible food. Fire has changed our ability to thrive. But even when technology does seem to have had an impact on human nature, it has taken hundreds or thousands of years for that change to be completed. Nuclear weapons have only been with us for a little more than seventy years. It seems highly unlikely, based on the evidence of human history, that nuclear weapons could have changed our nature so completely in so little time.

What strikes me most strongly, however, is the way this argument turns its back on the pessimism that is supposed to characterize realism. Genuine realists quote Hobbes with relish: Life, they say with a sort of grim satisfaction, is "solitary, poor, nasty, brutish, and short." Realists don't indulge in flights of fancy. They are supposed to have their feet firmly rooted in the hard rocks of fact. It is idealists who optimistically say that we can change the world by simply changing our hearts. Idealists believe that changing human nature is possible. Optimism is one of their defining characteristics.[15]

In the 1960s, a short, simple book about love called *Jonathan Livingston Seagull* appeared. Many read it, and many (in those heady days of social change) were deeply moved by it. Belief in change was in the air, and people were inspired to change their lives. They (and other young people inspired by other books) believed that a new society could be created, a utopian world community that would value love above all other things. And with this new emphasis on love, there would naturally come a world filled with peace. And we would all hold hands and sing.

My point here is not to mock those young people. The reason *Jonathan Livingston Seagull* matters to this argument is that it illustrates something about the claims of nuclear weapons advocates. What's striking about the optimistic faith in the ability to change human nature that *Jonathan Livingston Seagull* epitomizes is that nuclear "realists" are making essentially the same claim those hippies made. Nuclear believers say that the urge to make savage war has been overcome. They say we can now live in peace forever. There will be no more all-out wars. And they say this utopia of peace has already arrived (just without the singing). They're not claiming that it is a sweet and inspiring book that has changed human nature; they claim it is a tool—a piece of technology—that has brought about this magical transformation. But the underlying claim is the same: Human nature has been fundamentally altered, and what was constant is now gone forever.

But nuclear weapons have not magically transformed our warlike natures into calm and peaceful ones. Unbridled war, fought with savage abandon, is still likely, perhaps even inevitable. If you doubt that anger and violence are stalking the world, read some headlines. Around the world there are sudden fires of passion that leap up—first here, then there. War is raging in Europe as I write. With so much anger around as fuel, is there much doubt that a war that engulfs many nations, and many peoples is far off? The belief that large-scale war has been banished forever is nothing more than dangerous fantasy. All the evidence of history and everything we know about ourselves tells us that our warlike natures cannot change overnight. (That is the sound of genuine realism talking.)

Claims that we can change human nature are unsurprising in the mouths of gentle, pot-smoking hippies. On the lips of nuclear weapons proponents, they are realist heresy. The fact that nuclear weapons advocates can call themselves realists and at the same time claim that nuclear weapons make all-out wars impossible shows that they do not understand the assumptions that underlie their own position. Their "realism" is nothing of the kind.

* * *

All this talk about the inevitability of war is harsh stuff and reflects a relatively bleak view of the world (although in my defense, I am a man who looked into the nuclear abyss at six). People are not all bad. They are not without redeeming qualities. Progress toward a better world is possible. Human beings are an extraordinary mix of the highest and most inspiring idealism and the lowest and most appalling depravity. Both are present in any society and sometimes even in the same individual. The point here is that we cannot hope to deal with nuclear weapons unless we first take account of the realities of the world. Robert F. Kennedy wrote that even while undertaking an idealistic task, "If we are to act effectively, we must deal with the world as it is."[16] That sometimes leads to unpleasant and difficult conclusions.

But taking a darker and more pessimistic view of human nature also allows us to see that the stance of the people who want to keep nuclear weapons forever is flawed. And an understanding of their positions—the strengths and weaknesses—will arm us with the clearer view necessary to explain that elimination is not only possible but imperative.

* * *

The seventy-plus years of failure on the part of the people who argued for eliminating nuclear weapons prove nothing. It was a contest that was rigged from the beginning. The advocates for nuclear weapons claimed that they were the only realists in the debate and that ensured that they would win every time. But they aren't realists. They aren't.

3   See chapter 1 of Paul Boyer, *By the Bomb's Early Light: American Thought and Culture at the Dawn of the Atomic Age* (New York: Pantheon, 1985), pp. 3–26.
4   Boyer, *By the Bomb's Early Light*, pp. 33–45.
5   "[T]he strategy of manipulating fear to build support for a political resolution of the atomic menace helped fix certain basic perceptions about the bomb deep in the American consciousness." Boyer, *By the Bomb's Early Light*, p. 75.
6   Vladislav M. Zubok, "Stalin and the Nuclear Age," in *Cold War Statesmen Confront the Bomb: Nuclear Diplomacy Since 1945*, ed. John Lewis Gaddis, Philip H. Gordon, Ernest R. May, and Jonathan Rosenberg (Oxford: Oxford University Press, 2005, 1999), p. 51.
7   And it appears that the Baruch Plan faced strong opposition in the United States as well. "The story is told that a well-known general who, with Mr. Baruch, had been 'putting teeth' into the Lilienthal Plan, said that the final draft was finished, 'Now we have made it so stiff that even the Russians won't be fool enough to fall for it.' Whether true in detail or not, there was certainly a strong military and political group who had no sympathy at all with the high-flown and idealistic phraseology of the Lilienthal-Oppenheimer Plan and publicly campaigned against it." P. M. S. Blackett, *Atomic Weapons and East-West Relations* (Cambridge: Cambridge University Press, 1956), p. 91.
8   See Justice Breyer's excellent and incisive argument in Steven Breyer, *Making Our Democracy Work: A Judge's View* (New York: Alfred A. Knopf, 2010).
9   Quoted in Francis J. Gavin, *Nuclear Statecraft: History and Strategy in America's Atomic Age* (Ithaca, New York: Cornell University Press, 2012), p. 158.
10   John Lewis Gaddis, *The Cold War: A New History* (New York: Penguin Press, 2005), p. 52.
11   "News conference, President John F. Kennedy," State Department Auditorium, Washington, D.C., March 21, 1963, https://www.jfklibrary.org/Research/Research-Aids/Ready-Reference/Press-Conferences/News-Conference-52.aspx (accessed May 24, 2023).
12   Norman Angell, *The Great Illusion: A Study of the Relation of Military Power in Nations to Their Economic and Social Advantage* (London: W. Heinemann, 1910). Angell claimed that he hadn't declared that war was obsolete, only that it was "futile." Nevertheless, it was particularly popular with pacifists and those who feared war. See also Steven Pinker, *The Better Angels of Our Nature: Why Violence Has Declined* (New York: Viking, 2011).
13   William James, "The Moral Equivalent of War," in *War: Studies from Psychology, Sociology, Anthropology*, ed. Leon Bramson and George Goethals (New York: Basic Books, 1964), p. 23.
14   Sigmund Freud, "Why War?" p. 77.
15   The ideas in this section grew out of my reading of a marvelous essay by Ralph Waldo Emerson titled "The Conservative." Emerson talks about conservatism and liberalism, measures the benefits of each, points out the flaws that each one harbors, and concludes that both are essential. Highly recommended (especially in these intemperate times). Ralph Waldo Emerson, *Emerson: Essays and Lectures: Nature: Addresses and Lectures / Essays: First and Second Series / Representative Men / English Traits / The Conduct of Life* (New York: The Library

of America, 1983).
16. Robert F. Kennedy, *To Seek a Newer World* (New York: Bantam Books, 1968), p. 233.

2. SOCIETY-WIDE MISTAKES

It is possible because establishments have been mistaken about the value of powerful and important weapons before.

One reason people believe what the experts tell them about nuclear weapons is that so many politicians, generals, journalists, and other thought leaders have endorsed these weapons. Leaders echo what the experts say and have been echoing them for more than seventy years. It raises a question: how could all those people have been wrong? How could, essentially, the entire establishment of a country—government officials, scholars and commentators, TV personalities, and pastors preaching from pulpits—be wrong? And not just the establishment of one country, but the establishments of multiple countries. Not just the United States, but the governments and establishments of Russia and China and France and India and Israel and so on.

It's a good question. Experts are sometimes wrong, but when so many experts and officials from so many countries have looked at nuclear weapons and basically agreed that they are the most important weapons ever, it seems like they must be right.

Is it possible that all those people have been wrong? Of course, it is. We know it's possible because history gives us an answer. In fact, history provides us with certainty. History tells us that it is possible for the conventional wisdom about an important, powerful weapon to be wrong—very wrong. It is possible for government officials, even after years of soul-searching and careful thought, to be mistaken. It is possible for bureaucrats and military strategists and whole societies of people in different countries around the world to be wrong. It is possible not because it is theoretically possible, not because we can imagine it happening, and not because I say it is possible. It is possible because it has already happened. It is possible because it is historical fact. Not only has it happened, it has happened repeatedly throughout history. It happened in the ancient world with chariots, it happened in the medieval world with mounted knights, and, most recently, it happened with battleships.

Because this is such an important point, because people have trouble believing that the experts might be wrong, let me tell you the story of battleships as a parable—or perhaps a warning—about how large groups of people over long periods of time can be wrong about a weapon.

DREADNOUGHTS

The story of battleships spans more than five hundred years, but the part of the story we're interested in comes near the end of that tale—from roughly 1900 onward. It is a narrative of misperception, misjudgment, and mistake that may, in the end, have caused one European power to lose a war and undermined another's status as the leading nation of the world. It is a story that reminds us that forgetting to focus on the utility of a weapon can threaten a country's long-term security.

As European powers established more and more colonies in the middle 1500s, ships became a crucial part of their national economies. Successful colonial commerce required not only fleets of commercial ships to transport goods but also warships to protect them. As wealth flowed out of those colonies throughout the 1600s, 1700s, and 1800s, the importance of navies grew. And the most important fighting ships in those navies were battleships. People's eyes were drawn to battleships. They were the biggest and most powerful vessels, they tended to have the latest technology—they provoked awe.

Battleships seemed to vibrate with a meaning that stretched far beyond their ordinary military capabilities. They represented national power—expressing the majesty and might of a nation; they somehow seduced people to love their nation more; and they frightened nations whose battleships were smaller. This ability to magnify feelings, this strange aura of more intense feelings allowed them to accomplish tasks that mere shells and guns could not. In the century after 1800, battleships were such icons of power that a single ship could coerce an entire nation. Send a battleship into the harbor of a smallish power, make your demands, and many times the nation would simply bow down—not really to the ship itself but more to the ghostly shroud of symbolic power that seemed to surround it. It was known as "gunboat diplomacy," and it was a standard tool in coercing distant, weaker nations. It made a good argument for keeping battleships.

In 1902, the British navy's Rear Admiral Lord Charles Beresford captured this symbolic power in a single phrase. In a speech he gave in New York, he said that "battleships are cheaper than battles." Lord Charles neatly summed up the belief that battleships were such potent symbols that their symbolism alone could deter or coerce adversaries. Using symbolic power, rather than fighting wars, cost no lives and required no treasure. U.S. President Theodore Roosevelt liked the sentiment so much that he quietly inserted it into his speeches.[17]

Their symbolic power was evident even in the little things. For example, when Japan sent an ultimatum to China's head of state in 1915, in order to subtly intimidate China's leaders, the diplomats who drafted the document had their demands printed on paper with a battleship watermark. Even the paper was a reminder of Japan's majesty and military might.[18] The symbolic value of battleships was everywhere in international relations throughout the eighteenth and nineteenth centuries.

The part of this story that matters to us comes just after the world crossed over the cusp from the nineteenth century into the twentieth century—in 1906, to be exact—with the launching in Great Britain of the *HMS Dreadnought*. This battleship touched off extraordinary emotions when it was christened and first put to sea. The *Dreadnought* was an improvement over previous types of battleships, with more powerful steam turbine engines and larger guns. But the reaction was out of all proportion to the rather slight increase in military value it represented.

The *HMS Dreadnought* was met with delirium in Great Britain. On its maiden voyage up the Thames, more than a million people lined the banks of that river to cheer. The area where it was docked had to be regularly closed off because crowds of up to thirty thousand people, hoping for a closer look, would pack the pier so tightly that they risked the lives of small children, the weak, and the elderly.

Because of its new, faster engines and its reliance on massive twelve-inch guns, the *Dreadnought* was seen as a "revolution" in battleship design. Soon the name "dread-nought" became the generic term for a class of new, heavier, faster warships modeled on Great Britain's breakthrough warship.

The reaction in Great Britain was remarkable.[19] "The influence of the *Dreadnought* spread far beyond the limits of strategic and diplomatic concerns. The ship played roles in national identity and imperial sentiment," according to one historian of the period. Commercial products adopted the *Dreadnought* name. Songs, books, and poems celebrated the ship as an emblem of British greatness.[20] Eventually, the ship acquired "an almost mystical, lineal connection to a British past infused with ideas of justice and liberty."[21]

This overwrought response to the *Dreadnought* was echoed throughout Europe and across the globe. But unlike the joy and celebration in Britain, in other countries the *Dreadnought* inspired apprehension, alarm, and envy.[22] From Washington to Moscow, Tokyo to Santiago, government ministers told their masters that it was essential for the national security of their country to acquire these new vessels. Dreadnoughts, they explained, were necessary for maintaining military strength, diplomatic prestige, and national pride. Described as "the most deadly fighting machine ever launched in the history of the world,"[23] the *Dreadnought* touched off a massive arms race.[24]

> The Japanese, fresh from their triumph at Tsushima, responded almost immediately to the *Dreadnought* with their own *Satsuma* and continued building all-big-gun ships as rapidly as their limited industrial base would allow. The French, though no longer much interested in sea power, could not long resist the trend, and by 1911 the first true Gallic dreadnought, *Jean Bart*, was sliding down the ways. In 1910 the Italians added the nineteen-thousand-ton *Dante Alighieri* and kept on building. The Russians, after losing virtually their entire fleet in 1905, were naturally susceptible to dreadnought fever, and laid down the Pervoz-vannyi class as early as 1906 and then the Gangut class in 1911. Even Austria-Hungary, with precious little seacoast to defend, built the twenty-thousand-ton *Viribus Unitis*. Those that could not build dreadnoughts of their own, like Turkey and Brazil, simply ordered them built by others.[25]

Similarly, in the United States, a major shipbuilding effort got underway. This effort seemed so important, and "[t]he relationship between a strong fleet and a successful foreign policy was so vital" in President Teddy Roosevelt's eyes, that the president "assumed personal control over the service, serving as his own de facto naval secretary through his entire administration."[26]

But perhaps the most significant decision made about dreadnoughts was made in the capital of Britain's chief rival—Germany. Germany was by nature a land power, but Kaiser Wilhelm II decided that Germany's traditional dominance on land should be augmented with a challenge to British dominance at sea. Wilhelm was Queen Victoria's grandson; he had spent many summers in England mixing with British aristocracy and racing his yachts against theirs. He admired (and probably envied) Britain's navy. Battleships, he felt, represented a special kind of national power.[27] The Kaiser once told the King of Italy, "All the years of my reign my colleagues, the Monarchs of Europe, have paid no attention to what I have to say. Soon, with my great Navy to endorse my words, they will be more respectful."[28] The Kaiser didn't talk in terms of the military benefit that battleships would bring him. He talked about respect. Germany embarked on a massive dreadnought-building campaign, and by 1917, they had sixteen of these imposing, expensive dreadnought-class battleships.

The British, seeing the activity in German shipbuilding yards, came to believe that the new German ships were a threat to their dominance of the world's seas, and they accelerated their own building of dreadnought-class warships. By 1917, barely eleven years after the first battleship of this kind had been launched, Great Britain had more than thirty of these giant battleships. This arms race, between a rising power in Europe (Germany) and the dominant power in the world (Great Britain), heightened tensions between the two countries, fomented suspicion, and in the end shaped the course of events for decades to come.

World War I was the largest war that had ever been fought to that point in human history. It involved thirty-two nations, its effects spanned the entire globe, and it ultimately led to an estimated nineteen million soldiers and civilians killed. But even though they were seen as critical, battleships played almost no serious military role in the fighting. One major sea battle, Jutland, was fought. But the outcome was inconclusive, and neither side took heavy losses.

The fact that battleships played such a small role in the actual fighting might lead you to guess that afterward, people rethought their importance. But that didn't happen. The hold that battleships had on the hearts and imaginations of government officials and military officers around the world was so strong that in the years after World War I, the fascination with battleships continued unabated.

In the interwar years, battleships' influence spread beyond war to the pursuit of peace. In a number of nations, large and politically influential peace movements arose during the 1920s. Shocked by the worldwide scope of the war, horrified by the brutality of trench warfare, and appalled by the casualties, people were determined to prevent war from breaking out again. Some argued that only teaching nonviolence or supporting the League of Nations could prevent war, but the most popular plan for preventing war was focused on battleships. Having watched Germany and Great Britain pour untold national treasure into the dreadnought arms race, and having seen the tensions that that competition caused, they believed that battleships had been pivotal in causing World War I. The key to preventing war, they concluded, was to prevent arms races.

IT IS POSSIBLE

Officials in government agreed with this analysis, apparently, and in the years between World War I and World War II, enormous diplomatic effort was poured into negotiating treaties that limited the number, size, and kind of battleships that could be built. The Washington Treaty of 1922 and the two London Treaties that followed in 1930 and 1936 made limits on battleships the centerpiece of their terms. The treaties were hailed as important diplomatic accomplishments.

But of course, the treaties didn't prevent war. When war came again in 1939, it came for reasons that had nothing to do with battleships. And perhaps even more surprising, as the fighting swept across the globe, battleship after battleship was sunk. In the Atlantic, in the Mediterranean, in the Pacific, and in every other place where battleships were caught without air cover in World War II, they were pummeled and sunk almost without exception. It turned out that even though battleships were impressively large and carried the latest in big guns, weapons had been developed that made them stunningly vulnerable.

Three new technologies spelled the end of the battleship's dominance. The first was torpedo boats—small, very fast, and inexpensive boats powered by new engines and armed with torpedoes. They made it possible to attack battleships in swarms and sometimes sink them.

Torpedo boats began to appear even before the *HMS Dreadnought* was built. And although they initially lacked the range to fight on the oceans, it should have been clear that they posed a significant challenge to battleships.[29] Already in World War I, British battleships feared sailing too close to the coast of Germany because of torpedo boats. Denying battleships the ability to sail safely in coastal waters was a significant reduction in their military utility. And torpedo boats were far more economical—both in terms of cost to build and lives lost if they sank—than giant, expensive battleships.

The second new weapon that spelled obsolescence for battleships was also present when the Dreadnought hysteria first took hold: the submarine.[30] Submarines could approach battleships by stealth and sink them before anyone knew that danger was at hand. Although submarines were not yet as capable and effective as they would eventually be, they were already feared. As one historian noted, "[The British] Grand Fleet itself declined to venture into the central North Sea after 1916, such was its respect for the increasingly effective German U-boats."[31]

Even as early as 1906, it should have been obvious to an objective observer that torpedo boats and submarines posed a deadly challenge to battleships. Because of their enormous expense, battleships were too valuable to be exposed to danger from smaller, cheaper ships. And British

SOCIETY-WIDE MISTAKES

military and government officials at the highest level were aware that submarines posed new and very serious problems.[32] Yet somehow the powerful emotions that battleships evoked blinded them to battleships' coming obsolescence.

It was the arrival of World War II that finally laid bare the harsh reality. In that conflict, submarines and torpedo boats attacked battleships (much to the dismay of admiralties, especially the British). But the real threat came from a new weapon: the dive bomber. A battleship caught out on the open ocean by dive bombers was as good as sunk. The fate of the Japanese battleship *Yamato*—at 78,990 tons one of the heaviest battleships ever built—is representative. During the U.S. invasion of Okinawa in the early part of 1945, the *Yamato* sailed out of port to attack the Allied invasion fleet. Despite the fact that eight destroyers and a cruiser were providing it with anti-aircraft fire support, once the *Yamato* was sighted, it took less than two hours for U.S. dive bombers to sink her.

Battleships, at one time the linchpin of national security, were reduced by the end of World War II to little more than floating artillery platforms: useful for bombarding shore positions once full control of the air and sea had been established but unable to protect themselves in an open combat environment. World War II finally drove home the point that the day of the battleship was over. But it took two wars and scores of battleships being sunk before that reality finally sank in.

What was gained as a result of this remarkable infatuation with battleships? Or to put it another way, what did all the money and diplomatic effort poured out on battleships yield? Almost nothing. World War I had no massive sea battles like Tsushima Straits where battleships determined the outcome of the war.[33] The few battleships that saw actual combat won little glory—for the most part, they were sunk by mines, sunk by submarines, or unable to close with the enemy.[34]

The hard-won diplomatic treaties that so absorbed the diplomats of the world between the 1920s and 1930s did nothing to forestall World War II. Like shoveling sand into the rising tide, that diplomatic effort was as pointless as the original arms race had been.

Finally, what was the return on the money spent on battleships in the years between World War I and World War II? What happened to the dollars, marks, francs, yen, rubles, lire, and so on spent on giant iron ships between 1918 and 1936? It turned out to be good money thrown after bad. Battleships were rarely decisive in World War II, and they certainly did not play a role commensurate with their cost.

So, the planning, money, and hard negotiating invested in battleships were largely wasted. The powerful emotions that battleships evoked somehow deceived governments so much that—for decades—they were unable to see clearly the practical value (or lack of value) of battleships. But the inability of government officials to distinguish reality from emotion had a larger impact, especially on the two main rivals in the arms race, Germany and Great Britain. For Germany, it seems likely that their decision to build dreadnoughts was at least partly responsible for Great Britain joining World War I. As a result of their huge naval spending, Germany became fixed in the minds of British government officials as a dangerous threat. And that perception seems to have played a crucial role in tipping Great Britain's decision to come into the war. The Kaiser's fixation on dreadnoughts, therefore, was probably half the reason Germany lost the war. (The other half was the decision to launch unrestricted submarine warfare in 1917, a decision that eventually brought another adversary—the United States—into the war.) The inability of the Kaiser to see the reality rather than an emotion-infused image severely damaged Germany's long-term national security.

But perhaps the most severe consequences were meted out to Great Britain. Even though they ended up on the winning side in both wars, it is difficult to argue that the British decision to spend massively on battleships constituted a wise investment in their national security. British battleships saw almost no combat and played no meaningful role in the outcome of either World War I or World War II. And that massive spending had a large and negative long-term impact on British security. Economists and historians now argue that the runaway naval spending during these years played a significant part in depleting British finances and therefore in its fall from the greatest power on the earth to only one of many middle-sized powers.[35]

CONCLUSION

What the story of battleships shows is that it is possible for entire establishments or even whole societies to be wrong. Infatuation with an impressive weapon can affect anyone, from the lowliest academic scholar or military planner to the most exalted government minister or admiral of the fleet, from an experienced foreign policy journalist to the average person on the street. These misperceptions, once enough people have signed on, seem to take on a momentum of their own. The battleship myth overcame differences of culture, doctrine, and habit of mind. People in Europe, Asia, the United States, and in Latin America were all carried away by the frenzy for dreadnoughts. And once their eyes were fixed on battleships, it was hard for experience to cure their strange tunnel vision. Decades went by and evidence accumulated, yet still the faith in battleships persisted. Only the harshest reality—the humiliating loss during World War II of these expensive warships again and again at the hands of smaller, cheaper weapons—eventually broke through. What the story of battleships teaches us is that it is possible for a lot of people to be very wrong about the value of a weapon for a very long time.

* * *

So, what have we learned? What we've learned is that, although we should be respectful of expertise and established beliefs, that doesn't mean we can't also recheck their math and apply skepticism to what they say. Human beings are fallible creatures, and somehow weapons that become symbols exert an especially strong pull on human perceptions.

If anyone asks whether it is possible that so many well-intentioned, well-informed officials and experts could be wrong, we can say, with absolute confidence, "of course it is possible." It is possible for societies to be wrong about a weapon for decades. If anyone was inclined to doubt that conclusion, the story of battleships removes all doubt.

If experts and government officials were wrong about battleships at the turn of the twentieth century, then it is possible that they are wrong about nuclear weapons today. Even though there are seventy years of accumulated thinking about nuclear weapons, we should not necessarily feel bound by it. Questioning the assumptions of the last seventy years is not a fool's errand. It may well be essential work. Maybe nuclear weapons are not necessary. Maybe it is possible to eliminate these dangerous weapons sometime soon. The claims of nuclear weapons advocates, even though they have expensive educations and carefully burnished credentials, could still be wrong.

People who think that generations of experts in countries around the world saying that nuclear weapons are essential weapons proves that they are, in fact, essential need to study a little history. And all the rest of us, who think that the questions about nuclear weapons are settled, who think the best course is to put the matter out of their heads, need to remember that this matter is not settled. It is possible that the experts and officials—even though they have all agreed for decades—have made fundamental errors in their thinking about nuclear weapons.

17 Robert K. Massie, *Dreadnought: Britain, Germany, and the Coming of the Great War* (New York: Random House, 1991), p. 514.
18 Margaret MacMillan, *Paris 1919: Six Months that Changed the World* (New York: Random House Trade Paperback, 2001), p. 328. Watermarks are variations in the thickness of paper so that when you hold the paper up to the light, a pattern or image is visible.
19 See the excellent chapter by Jan Rüger on the symbolism of dreadnoughts. Jan Rüger, "The Symbolic Value of the Dreadnought," in *The Dreadnought and the Edwardian Age*, ed. Robert J. Blyth, Andrew Lambert, and Jan Rüger (Farnham, Surrey: Ashgate, 2001), pp. 9–18.
20 "When the Dreadnought was launched, it was embraced by popular culture almost instantly. There were dreadnought songs, poems, books, and films." Rüger, "The Symbolic Value of the Dreadnought," p. 9.
21 Robert J. Blyth, "Introduction," in *The Dreadnought and the Edwardian Age*, ed. Robert J. Blyth, Andrew Lambert, and Jan Rüger (Farnham, Surrey: Ashgate, 2001), p. 4.
22 As the scholar Michael Howard explains, at that time powerful navies were "a status symbol of universal validity which no nation conscious of its identity could afford to do without." Quoted in Michelle Murray, "Identity, Insecurity, and Great Power Politics: The Tragedy of German Naval Ambition Before the First World War," *Security Studies*, no. 19 (2010), p. 675.
23 One of the important reasons that the military utility of dreadnoughts did not match the hype is that the sighting technology available at the beginning of the twentieth century was insufficient to make those massive guns effective. They could fire long distances, but it was very difficult to hit anything. See Robert L. O'Connell, *Sacred Vessels: The Cult of the Battleship and the Rise of the U.S. Navy* (New York: Oxford University Press, 1991), pp. 145–146.
24 Giles Edwards, "How the Dreadnought Sparked the 20th Century's First Arms Race," *BBC News Magazine*, June 2, 2014. http://www.bbc.com/news/magazine-27641717. "The acceleration crisis of 1909 illustrates the 'cold war' dynamic in Anglo-German relations, with the naval arms race at its core." T. G. Otte, "Grey Ambassador: The Dreadnought and British Foreign Policy," in *The Dreadnought and the Edwardian Age*, ed. Robert J. Blyth, Andrew Lambert, and Jan Rüger (Farnham, Surrey: Ashgate, 2001), p. 51.
25 Robert L. O'Connell, *Of Arms and Men: A History of War, Weapons, and Aggression* (New York: Oxford University Press, 1989), p. 228.
26 Robert L. O'Connell, *Sacred Vessels*, p. 104.
27 "Germany's naval program was designed not for strategic reasons, but to secure recognition of its identity as a world power." Michelle Murray, "Identity, Insecurity, and Great Power Politics: The Tragedy of German Naval Ambition Before the First World War," *Security Studies*, no. 19 (2010), p. 658. See also Jan Rüger, "The Symbolic Value of the Dreadnought," pp. 9–18.
28 O'Connell, *Of Arms and Men*, p. 227.
29 The utility of these small, fast, inexpensive boats was so obvious that George Leo von Caprivi, commander of the German navy from 1883 to 1888, chose

to focus all of Germany's naval budget on torpedo boats. These quick, deadly craft were the ideal weapons for defending Germany's coasts. When Kaiser Wilhelm succeeded to the German throne in 1888, however, this limited aim was abandoned. Wilhelm was determined to build "big boats" and accepted Caprivi's resignation. Massie, *Dreadnought*, pp. 162–163.

[30] This knowledge was growing in the British high command. "Fisher believed that the new gyroscope torpedoes undermined the utility of the battleship. Battleships, he argued, had been essential when they were the only things that could sink other battleships. The 'ultimate security of the defense' had now been lost; battleships could be sunk by torpedo craft, surface and sub-surface." Eric Grove, "The Battleship Dreadnought: Technological, Economic, and Strategic Contexts," in *The Dreadnought and the Edwardian Age*, ed. Robert J. Blyth, Andrew Lambert, and Jan Rüger (Farnham, Surrey: Ashgate, 2001), p. 172.

[31] Paul Kennedy, "HMS Dreadnought and the Tides of History," in *The Dreadnought and the Edwardian Age*, ed. Robert J. Blyth, Andrew Lambert, and Jan Rüger (Farnham, Surrey: Ashgate, 2001), p. 223.

[32] The danger posed by submarines was frightfully plain to British naval officials. Sir John Fisher, First Sea Lord from 1904 to 1910, summarized the problem in a memorandum to the First Lord of the Admiralty, Winston Churchill. "[The submarine] cannot capture the merchant ship; she has no spare hands to put a prize crew on board; little or nothing would be gained by disabling her engines or propeller; she cannot convoy her into harbor; and, in fact, it is impossible for the submarine to deal with commerce in the light and provisions of accepted internal law.... There is nothing else the submarine can do except sink her.... This submarine menace is a terrible one for British commerce and Great Britain alike, for no means can be suggested at present of meeting it except for reprisals." Robert J. Art, "The Influence of Foreign Policy on Seapower: New Weapons and Weltpolitik in Wilhelminian Germany," in *The Use of Force: International Politics and Foreign Policy*, ed. Robert J. Art and Kenneth N. Waltz (New York: University Press of America, 1983), p. 173.

[33] The Battle of Jutland, the only real clash between the British Navy and German Navy, was largely inconclusive and ended before all the forces were engaged.

[34] See the description of the six warships damaged during the Dardanelles campaign in Kennedy, "HMS Dreadnought and the Tides of History," p. 225.

[35] "[T]he British Treasury felt itself under heavy pressure in the two decades before 1914 as the newer technology more than doubled the price of an individual battleship." Paul Kennedy, *The Rise and Fall of the Great Powers: Economic Change and Military Conflict from 1500 to 2000* (New York: Vintage Books, 1989), p. 230.

3. IN THE BEGINNING

*It is possible because what experts
think they know is probably wrong.*

Could it be that people today are gripped by the same confusion that wasted so much money and effort between 1900 and 1945? Are we mistakenly infatuated with nuclear weapons in the same way people in the early twentieth century were infatuated with battleships? Is it possible that we don't see nuclear weapons realistically?

EMOTIONS

To answer these questions, we could carefully go back over the history and double-check to make sure the facts align with our ideas. We could review the theories and look for logical inconsistencies. Both approaches have been used in nuclear weapons scholarship over the last sixty years. But what the most recent research in psychology and neuroscience seems to show is that even the most rational thinking is intertwined with feelings; that emotions are inextricably linked with almost everything we think or do. For decades, though, emotion has been largely excluded from the formal debate about nuclear weapons. Much of the nuclear weapons literature leaves the impression that nuclear weapons exist in a world where emotions don't even exist, where killing millions of civilians, for example, is called "collateral damage" and decisions during a crisis are made based on calculations or logic equations. Most scholarly papers in the field often read as if fear, the desire for revenge, foolish optimism, lust for war, or any of the other emotions that so often control human actions don't even exist.

Given the recent research that shows how central emotions are to decisions and actions, perhaps it makes sense to start by reviewing some of the emotions that might have shaped the first ideas and conclusions about nuclear weapons. So, rather than examining policy statements or academic theories, rather than digging into the facts (although we will eventually get to the facts), let's begin with emotions. The latest research says that to understand the reality of human behavior, we need to look at emotions. So, let's do that.

For this part of the investigation, we're going to concentrate on the emotional reactions in the United States. Nuclear weapons were first developed in the United States,[36] the reality of what it meant to use and possess the weapons was first confronted in the United States, and those first reactions seem to have informed much of the thinking that other countries eventually adopted toward nuclear weapons.[37] So, with apologies, we will turn first to the United States.

CREATION

Let us begin at the beginning. The year is 1945. It is night in the high desert of New Mexico. The weather is bad, and wind whistles and whines through the metal struts of the tower that holds the "gadget." Thunderstorms have raged here for two nights in a row, and now lightning can be seen striking in the distance.

J. Robert Oppenheimer, the scientist in charge of the bomb's development (thin, somber, high-strung, and ambitious) and General Leslie R. Groves, the military engineer responsible for the operation (driven, humorless, overweight, self-important) emerge again from the south control bunker and look up at the clouded night sky. Will the test have to be scrapped? Twice already that night, bad weather forced the test to be postponed. Everyone is restless and impatient, waiting for the conditions to be right. One last check with the meteorologist and the test is set for 5:30am.

Groups of scientists and military personnel prepare themselves. They are in trenches and bunkers well back from the tower. They smear suntan lotion on their faces, wear special goggles, and lie face down, feet toward the bomb, on desert sand. One scientist, Enrico Fermi, has scraps of paper he intends to release when the bomb's shock wave arrives. He hopes he can make an immediate rough estimate of the bomb's strength by how far they are blown. Another, Otto Frisch, is telling himself to stay calm and memorize everything about the phenomenon as precisely and unemotionally as possible. Groves is wondering again if he's done everything necessary to evacuate in case of disaster. Oppenheimer goes back and forth between fear that the experiment will fail and fear that it will succeed.[38]

Anticipation hangs in the air. In one later account, William Laurence, a reporter for the *New York Times* brought in as an observer, "shivered in the cold and moved forward on the knoll now dotted with murmuring shadows. He was an atheist by nature, believing firmly in the inevitability of events. But even Laurence, like others on the hill, sensed he was about to share in a profound religious experience, an event bordering on the supernatural."[39]

And then the blast.

> The whole country was lighted by a searing light with the intensity many times that of the midday sun. It was golden purple, violet, gray and blue. It lighted every peak, crevasse and ridge of the nearby mountain range with a clarity and beauty … [that] the great poets dream about but describe most poorly and inadequately. Thirty seconds after the explosion came, first, the air blast pressing hard against the people and things to be followed almost immediately by the strong, sustained, awesome roar which warned of doomsday and made us feel that we puny things were blasphemous to dare tamper with the forces heretofore reserved to the Almighty.[40]

As the fireball and huge cloud of smoke rise higher and higher into the night, the scientist J. Carson Mark is suddenly seized with the irrational thought—another part of his brain struggles to convince him it cannot be true—that the tower of smoke and fire will keep rising and growing until it fills the entire sky.[41]

Read accounts of that pre-dawn moment and you can't help but notice how awestruck those men were. They grope for words strong enough to describe the experience. They remember it as an overwhelming and intimidating moment. Military men and scientists are not known for exaggeration or hyperbole, but it is striking how many of them, in trying to convey the profound power of the bomb, reach for images from religion. "A great blinding light lit up the sky and earth as if God himself had appeared among us…. [T]here came the report of an explosion, sudden and sharp as if the skies had cracked … a vision from the Book of Revelation."[42] Profound emotions seem to wrap around the entire event. Most of these people seemed to feel that there was something "supremely mysterious, majestic, almost divine, in [this] manifestation of atomic power."[43]

Perhaps what best illustrates this sense of the mysterious and the divine is the story of the blind woman. A young woman was being driven across the New Mexico desert by her brother early on the morning of July 16, 1945. Suddenly, a bright light flashed in the pre-dawn sky (the light from that first nuclear test) and the blind woman—startled—said, "What was that!?"[44] By itself, this is not a very remarkable story. People who lose their sight often have some residual visual perception, and being aware of especially bright lights is not uncommon. But the story is not told to illustrate medical facts or how bright the explosion was. What makes the story interesting is that it is repeated with a hushed tone and a nod to the miraculous. The bomb, the teller implies, is so powerful it can make a blind person see.

Oppenheimer, watching the fire and smoke rise, claims to have suddenly thought of the words from the sacred Hindu scripture, the Bhagavad Gita: "Now I am become Death, the Destroyer of Worlds."[45]

So, overpowering awe and also a sense of the divine. It is useful to note these emotions, not because their feelings necessarily reflect objective reality but because that's what the first people to witness a nuclear explosion felt.

FIRST REACTIONS

Second, let's widen the viewfinder a little and examine what people in the United States felt. We've seen what people present at the creation felt. How did the larger U.S. population react to the news that Japanese cities had been destroyed by atomic bombs? Curiously, reports that Hiroshima and Nagasaki had each been destroyed by a single bomb somehow seem to have filled many Americans with a sense of danger that was looming just over the horizon. The bombings, *The Christian Century* said, "cast a spell of dark foreboding over the spirit of humanity."[46] The *New York Herald Tribune* commented, "[One] forgets the effect on Japan ... as one senses the foundations of one's own universe trembling."[47] A letter to the editors of the *New York Times* remarked that events were spreading "a creeping feeling of apprehension" across the nation.[48] A correspondent for the *New York Sun* described the mood in Washington, D.C.: "For forty-eight hours now, the new bomb has been virtually the only topic of conversation and discussion... For two days it has been an unusual thing to see a smile among the throngs that crowd the streets. The entire city is pervaded by a ... sense of oppression."[49]

President Harry Truman, addressing the nation on the radio the day after the Nagasaki bomb was dropped, sounded chastened. "It is an awful responsibility which has come to us. We thank God that it has come to us instead of to our enemies; and we pray that He may guide us to use it in His ways and for His purposes."[50]

Even the announcement of victory did not seem to dispel the worry. *The New Republic* reported when Japan's surrender was announced a week later, it did not lift the anxious feeling brought by the initial news of the bombings. In Washington, D.C., it reported, there was a "curious new sense of insecurity, rather incongruous in the face of victory."[51] Edward R. Murrow summed it up this way: "Seldom, if ever, has a war ended leaving the victors with such a sense of uncertainty and fear, with such a realization that the future is obscure and that survival is not assured."[52]

Time magazine, in its first postwar issue, wrote:

> The knowledge of victory was as charged with sorrow and doubt as with joy and gratitude. In what they said and did, men are still, as in the aftershock of a great wound, bemused and only semi-articulate.... But in the dark depths of their minds and hearts, huge forms moved and silently arrayed themselves: Titans, arranging out of the chaos an age in which victory was already only the shout of a child in the street.[53]

You could say that people were worried, but their feelings seem to be stronger than simple worry. Again and again, they imagined not just danger but the end of everything—apocalypse. Hanson W. Baldwin, the *New York Times* military correspondent, opined in an edition of *Life* magazine devoted to the atomic bombings that as soon as the long-range rockets the Germans had developed could carry atom bombs, they would "destroy cities at one breath." Mankind, he said, had "unleashed a Frankenstein monster." Conventional war and the infantrymen who fought it were now obsolete.[54]

The NBC radio commentator H. V. Kaltenborn admonished his listeners: "As we listen to the newscast tonight, as we read our newspapers tomorrow, let us think of the mass murder which will come with World War III." Later he warned, "We are like children playing with a concentrated instrument of death whose destructive potential our little minds cannot grasp."[55]

THE SWORD

Finally, let's examine what happened once those emotions had crystallized. How did the emotions that were let loose into American culture eventually settle and harden? What mental images were left in people's minds? When they thought of the predicament they were in, what metaphors and similes did they use?

IT IS POSSIBLE

Fast forward from the end of World War II to 1961. The scene is the great hall of the United Nations General Assembly, crowded with representatives from around the world. They are turning and talking to one another in their seats, and there is expectancy in the air. A young man—younger than any other person ever elected to the office—is president of the United States. His speeches during the campaign and at his inauguration were inspiring, but already his inexperience seems to be showing. Last spring, he approved an invasion of Cuba by CIA-trained rebels which was quickly crushed, leaving the United States looking weak and foolish. And all through this summer he has been engaged in a dangerous struggle of wills with the premier of the Soviet Union over the divided city of Berlin. The United States has called up military forces and added spending to their already large defense budget. The Soviets have shifted forces and tightened access to the city. The days have been anxious and tense because, for months, the possibility of war—the very real possibility of nuclear war—has hung like a gathering storm over Europe, the United States, and the Soviet Union.

The young man is striding up to the dais now, and after a brief pause, he begins. He opens with regret for the recent death of the U.N. Secretary General, Dag Hammarskjöld but quickly moves to his larger theme: disarmament. The president talks about the dangers of war, the new necessity of disarmament, and the role that the United Nations must play. The tone is sober, challenging, and eloquent.

> For in the development of this organization rests the only true alternative to war, and war appeals no longer as a rational alternative. Unconditional war can no longer lead to unconditional victory. It can no longer serve to settle disputes. It can no longer be of concern to great powers alone. For a nuclear disaster, spread by winds and waters and fear, could well engulf the great and the small, the rich and the poor, the committed and the uncommitted alike.[56]

The purpose of the speech, it turns out, is to propose a ban on tests of nuclear weapons above ground. This proposal, which will grow into the Nuclear Test Ban Treaty, is the first important step in slowing the arms race. But before he gets to the heart of the speech, the young president prepares the way with an image of such extraordinary power and vividness, an image that so deftly captures the feeling of living under the threat of nuclear war, that it will be repeated and quoted for decades. He says:

> Today every inhabitant of this planet must contemplate the day when this planet may no longer be habitable. Every man,

woman and child lives under a nuclear sword of Damocles, hanging by the slenderest of threads, capable of being cut at any moment by accident or miscalculation or madness. The weapons of war must be abolished before they abolish us.[57]

"Every man, woman and child lives under a nuclear sword of Damocles." It is a powerful image of constant and deadly danger.

It refers to a famous story that has come down to use from an ancient kingdom in Sicily. King Dionysius was the ruler of the city of Syracuse in the fourth century BC, and Damocles was one of his courtiers. One day when he was praising the king's greatness and all the many fine things that being ruler of a wealthy community allowed him to do with as he chose, Dionysius surprised him by offering to change places with him for one day. Damocles quickly accepted. But he found when he was seated in the king's seat that, although there were luxuries and riches spread before him, there was also, dangling over his head, suspended by a single strand of horsehair, a razor-sharp sword. Dionysius was illustrating for his courtier that, although being a ruler has many benefits, it also comes with frightening dangers.

This young president, when he wished to sum up the situation that the world was living in said that we were all like Damocles: forced to live under a grave and deadly danger, delicately balanced, that could kill us at any time. But it wasn't just the image that is so interesting. It is the fact that it has resonated with Americans for so long. The fact that it continues to echo in people's heads suggests that even after people had been living with nuclear weapons for many years, they still remained deeply fearful about them.

COLD FEAR

What this exploration of feelings reveals is that nuclear weapons changed the way people felt. They made people think of the power of gods—gods who could destroy whole worlds. They made Americans fearful and depressed even in the face of victory. And the fear lasted, decade after decade, making people feel like there was a sword hanging over their heads.

Of course, over time it would be natural that these feelings, like almost all feelings, would fade. But instead of fading and eventually healing, they appear to have been transformed by two unusual and important circumstances—circumstances that seem to have turned them into permanent scars on the soul of American society. The first circumstance was a frightening confrontation between the United States and the Soviet Union, a sort of international shoving match that exacerbated people's feelings of fear for more than forty years. The second was an

IT IS POSSIBLE

unprecedented lack of experience. Together, these two factors appear to have heated and hardened what people were already feeling. Let's examine each in turn.

In the years after the end of World War II, the United States increasingly found itself at odds with one of its former allies: the Soviet Union. This confrontation never crossed over into actual war but continued on, year after year, in an odd sort of near-war. Eventually, this bitter clash that always stopped just short of war came to be called the Cold War. Normally, being in conflict with a distant land would not have caused Americans to fret or worry. But they were already unsettled by nuclear weapons, and the advent of long-range bombers and then rockets somehow made Americans deeply afraid.

The strength of the emotion was partly the result of the fact that it felt new and unexpected. For more than 150 years, the United States had thought of itself as so isolated as to be almost entirely safe. With wide oceans east and west and friendly neighbors north and south, for generations it had seemed that the country was invulnerable. For example, listen to this boast in a speech by a young Abraham Lincoln:

> All the armies of Europe, Asia and Africa combined, with all the treasure of the earth ... with a Buonaparte for a commander, could not [attack the United States and get far enough to] take a drink from the Ohio [River] ... in ... a thousand years.[58]

The advent of nuclear weapons, combined with the arrival of the Cold War, seemed to rip that sense of security away in an instant. Suddenly, inexplicably, people in the United States found themselves vulnerable to devastating attack. For more than a century, they had been living far from the dangers of Europe and the rest of the world, focused on building and making, and suddenly awoke one morning to find themselves strapped onto a bull's eye. They went from the comfort of complete security to naked vulnerability, seemingly overnight.

The depth of the fear they felt can be measured by the strange things they did. Take, for example, the McCarthy "witch hunts." Senator Joseph McCarthy achieved national prominence by repeatedly waving a sheaf of papers and declaring that it contained a list of known communists, spies, and homosexuals working within the State Department, the Army, the CIA, and other institutions of U.S. government. Of course, he had no such list of names and no actual proof of anything. But he was unscrupulous, and his charges made front-page news across the country. His frightening accusations resonated with the feelings of the times, and millions of ordinary Americans believed him. McCarthy soon set his Senate subcommittee to work holding hearings to investigate—searching high and low for the enemies he was sure were hiding in government jobs. The accusations and speculative attacks McCarthy launched wrecked people's careers and even led some people to commit suicide. How could Americans—for so long such a practical, confident, optimistic society—have been fooled by his wild accusations? Fear, it seems, was what suspended people's judgment. In the end, McCarthy's fruitless hunt for spies and other "enemies" distracted the nation and interfered with the proper functioning of government for years.[59]

The fear that stalked U.S. society in those early years can also be discerned in one of the strangest educational decisions in the history of the United States. Serious educators, responsible government officials, and otherwise sensible parents decided that the best way to prepare young people for life in America was to have millions of children repeatedly crouch under their desks and imagine that it was the first moments of a nuclear war. Across the country, students and teachers held practice drills for what to do in the event of a sudden nuclear attack. In hindsight, these "duck and cover" drills were, as a practical matter, both obviously pointless and emotionally harmful. Do children need special training to ensure that they will listen to their teacher in a crisis? And how much actual protection will a school desk provide in a nuclear blast? The only real outcome of these drills was to emotionally scar untold numbers of children for life.[60]

It is hard to overstate how important the fear that nuclear weapons created was. Historian Paul Boyer argues that nuclear weapons were the central emotional reality of the Cold War.

> Years before the world's nuclear arsenals made such a holocaust likely or even possible, the prospect of global annihilation already filled the national consciousness. This awareness and the bone-deep fear it engendered are the fundamental psychological realities underlying the broader intellectual and cultural responses of this period.[61]

The Cold War amplified the fear that nuclear weapons touched off. It made the possibility of a civilization-destroying catastrophe seem real.

* * *

The reason all these long-ago emotions matter is that emotion can sometimes cloud judgment. Hysterical people often make poor choices. No one does their best thinking when they're afraid, which means that the fear and awe that people felt when confronted with nuclear weapons could well have led them to make mistakes. It is possible, perhaps even likely, that those early emotions led the people who were trying to understand nuclear weapons—to work out where they fit into reality—to make mistakes. Not because they were foolish or had bad intentions but because they were human.

A LACK OF FACTS

But, of course, any mistakes that might have been made happened a long time ago. Even if those emotions somehow led to errors of judgment about nuclear weapons, it stands to reason that those errors have been corrected at some point in the many decades that lie between that time and this. It has, after all, been more than three-quarters of a century since the first atomic explosion. Surely that is time enough for the experts and officials in charge to see these dangerous weapons with clear-eyed realism, isn't it?

The problem is that once a theory has taken hold, once a community of people gets used to thinking about things in a certain way, using certain assumptions, they tend to keep on thinking in that way. Unless some contradiction or conflict between the theory and facts forces them to reevaluate, they just stick with what they've been taught. Rethinking a theory usually only happens when a theory bumps up against the harsh, abrasive surface of reality. Unless reality reveals painful contradictions and inconsistencies in a theory's power to explain, people usually leave well enough alone. The difficulty here and the reason why mistakes might have persisted since the earliest days of the Cold War is that this method of discovering flaws is not available when it comes to nuclear weapons.

Nuclear weapons are remarkable in many ways: the enormous size of their explosions, their apparent ability to coerce surrender in just four days, the way they unleash the elemental forces of nature, and their giant, looming image in our minds' eye. But these things are not what really sets them apart, what makes them unique in history. What makes them different is how little practical experience we have with them. Most weapons are used again and again. Over time their properties and best uses become clear. Nuclear weapons are not like that. There has never been a major weapon where so much of what we think we know is theory.

Most weapons have been used repeatedly, in different situations, across different types of terrain, in all sorts of weather, by different troops, trying to achieve different military objectives, and so on. Experience—the use of a weapon again and again—is what makes an accurate assessment of a weapon's military value possible. The greater the amount of experience with a weapon, the greater the accuracy of the assessment.

But this is exactly what we don't have when it comes to nuclear weapons. Nuclear weapons were used just twice, in a single week, against a single type of target, against a single opponent, in a single war, and never used again. This is an enormous obstacle. How can we double-check our ideas about them? It's hard to get something right with just one or two uses. Imagine writing the manual for using a new operating technique—perhaps involving a new medical device—after using it in only one operation on one patient. Imagine writing the warnings for a new and powerful kind of chainsaw—one that bucks and kicks dangerously—after cutting down only one tree. Imagine writing a sociology doctoral dissertation based on only one or two facts.

Think about how little experience we have when it comes to nuclear weapons. They have not been used against different types of targets (airports, ships at sea, hardened facilities, and so on); they have not been used on different types of terrain (jungle, steep ravines, savanna, and so on); they have not been used in different kinds of wars (civil wars, revolutionary wars, wars of conquest, wars defending borders, and so on); they have not been used against different types of governments (democracies, dictatorships, oligarchies, and so on), against different kinds of troops (volunteers, veterans, religious or ideological fanatics, mercenaries, and the like), under different weather conditions (low overcast, torrential rain, heavy snowfall, and so on), or against different unit types (armored, air mobile, coastal patrol boats, dug-in artillery, headquarters, screening troops, and the like). Actually, they haven't been used on a battlefield—ever.

And this critical lack of information has not been filled in during the years since the end of World War II. No matter how many controlled tests—above ground, below ground, at sea, in space—have been conducted, they cannot make up for the lack of practical wartime experience.

This is unnerving. We have enormously powerful weapons, we have constructed much of the world's security system around them, and yet we have very little experience with which to evaluate their actual military utility. How serious of a problem is this likely to be?

To see how critical practical experience is when it comes to judging the military value of a weapon, consider the story of heavy bombers. In the years between World War I and World War II, countries were just beginning to build bombers big enough and sturdy enough to carry large payloads over long distances. These new long-range bombers posed novel questions. How useful would they be? Would bombers make a difference in future wars? No one had ever carried out large-scale bombing attacks on cities, so most government officials and soldiers had no answers. But some strategists felt sure they could predict the effects large bombing raids would have. After all, they had seen the reports about how destructive test bombing runs were. If an adversary's cities were heavily bombed, they confidently asserted, that country would be forced to surrender almost immediately. Giulio Douhet, one of the most famous and influential of these strategists, wrote:

> And if on the second day another ten, twenty, or fifty cities were bombed, who could keep all those lost, panic-stricken people from fleeing to the open countryside to escape this terror from the air? A complete breakdown of the social structure cannot but take place in a country subjected to this kind of merciless pounding from the air. The time would soon come when, to put an end to horror and suffering, the people themselves, driven by the instinct of self-preservation, would rise up and demand an end to the war—this before their army and navy had time to mobilise at all![62]

Douhet and other strategists were certain: bombing would be decisive. After all, hadn't the test structures been obliterated on proving grounds? So, governments invested heavily in bombers and used those bombers to launch campaigns of city destruction once World War II was in motion. But despite the confidence of Douhet and others, and despite extensive testing, when bombers were sent to bomb actual cities, the predictions that they would be decisive turned out to be wrong. The strategists had overestimated both the impact bombing would have on civilians and the speed with which damage to buildings and facilities could be repaired.[63] Instead of delivering decisive blows, the chief result of bombing cities was to lose bombers, kill a staggering number of airmen, and exponentially increase the suffering of civilians who had little or nothing to do with the war effort. The dean of nuclear strategists, Bernard Brodie, after studying the city bombing conducted during World War II, concluded that the bombing of cities was "an inordinate waste of bombs and of bombing effort."[64]

Strategists and military experts had tested bombers in practice runs, they had measured the size of the craters, blown up test structures, and extrapolated the scale of the damage that could be done. But it wasn't until bombers were actually used in war that reality came into focus. Predictions based on proving ground tests, it turned out, could be glaringly wrong. Battlefield experience was the only reliable way to accurately assess the value of military weapons.

Nuclear weapons have been tested thousands of times. The U.S. military has measured the shockwave, the fires, the EMP (electromagnetic pulse), the radiation, and the fallout again and again. It has built test houses, planted test forests, hauled in test vehicles, blown them up again and again, and carefully measured the results. This extensive testing has provided a moderately good understanding of the physics of nuclear explosions. (It has also provided a particularly large and unwelcome body of data about the effects of radioactive fallout and the aggressive cancers that people downwind can suffer as a result of those tests.) But as the story of bombers makes plain, the physics hardly tells the whole story. All these tests do not compensate for our lack of practical wartime experience.

EMOTIONAL ECHOES

Given the way nuclear weapons burst into the consciousness of people in the United States, it would not be surprising if they had misjudged those weapons. The fear and awe that nuclear weapons seemed to trigger in people were only intensified by the dangers of the Cold War. And without any practical experience with the weapons, there was no way for people to get their feet firmly back on the ground. Everything seemed to be intense emotion and speculation.

So, if there was a certain tendency to exaggerate, a slightly hysterical tone to what was said and done, that would make sense. And it is relatively easy to find statements from that time that do seem overwrought. Here, for example, is David Lilienthal, who, as chair the Atomic Energy Commission from 1946 to 1950, had a front-row seat at the nuclear policymaking establishment. He's writing here in the early 1960s, reflecting on his experience with nuclear weapons (which he calls "the Atom"):

> The Atom has had us bewitched. It was so gigantic, so terrible, so beyond the power of imagination to embrace, that it seemed to be the ultimate fact. It would either destroy us all or it would bring about the millennium. It was the final secret of Nature, greater by far than man himself, and it was, it seemed, invulnerable to the ordinary processes of life, the processes of growth, decay, change. Our obsession with the Atom led us to assign to it a separate and unique status in the world. So greatly did it seem to transcend the ordinary affairs of men that we shut it out of those affairs altogether; or rather, tried to create a separate world, a world of the Atom.[65]

Nuclear weapons policy was born in a time of anxiety, fear, and even paranoia. You can feel the emotions from that time rising off the page. And there has been no practical experience since that time that could correct or refine those early ideas, no contact with reality that could draw attention to any mistakes that might have been made in those early Cold War years.

What that means is that those early theories about nuclear weapons have never been adequately tested. That is something that we should all be grateful for: no one wants to learn about the specifics of fighting a nuclear war. But it does leave us with a problem when we try to assess whether those first ideas formed about nuclear weapons were accurate. Do we understand the reality of nuclear weapons? Are we sure about the theories developed in the 1950s and never really put to the test since?

If it seems likely that mistakes were made in those first years, there is an urgent need to revisit these questions. Errors in thinking about matters that are this consequential and this important cannot go uncorrected. It is necessary to go back over the original ground and reassess the assumptions and theories that now lie at the heart of nuclear weapons policy.

36   Although the brilliant scientists who did the actual work came from many different countries.
37   Although there has not been unanimity among the nine nuclear-armed states. China, in particular, seems to have viewed nuclear weapons with real skepticism during the first sixty years of their possession of these weapons.
38   Robert Jungk, *Brighter Than a Thousand Suns* (New York: Harcourt Brace, Jovanovich, 1958), p. 200.
39   Lansing Lamont, *Day of Trinity* (New York: Atheneum, 1965), p. 225.
40   Richard B. Frank, *Downfall: The End of the Imperial Japanese Empire* (New York: Penguin Books, 1999), p. 261.
41   Lansing, *Day of Trinity*, pp. 200–201.
42   Ronald W. Clark, *The Greatest Power on Earth: The International Race for Nuclear Supremacy* (New York: Harper & Row, 1980), p. 199.
43   Spencer R. Weart, "The Heyday of Myth and Cliché," in *Assessing the Nuclear Age: Selections from the Bulletin of the Atomic Scientists*, ed. Led Ackland and Steven McGuire (Chicago: Educational Foundation for Nuclear Science, 1986), p. 82.
44   Boyer, *By the Bomb's Early Light*, p. 6.
45   Oppenheimer's famous quotation is a questionable translation. The word kala, is generally translated as "time," which would make the phrase: "Now I am become Time, the Destroyer of Worlds." Oppenheimer's version, which translates kala as "death," a much less rarely used meaning, makes the phrase much more apocalyptic. It is not clear whether Oppenheimer misremembered the phrase or whether the version he read contained this doubtful translation. M. V. Ramana, "The Bomb of the Blue God," *South Asia Magazine for Action and Reflection*, no. 13 (2001).
46   Boyer, *By the Bomb's Early Light*, p. 6.
47   Boyer, *By the Bomb's Early Light*, p. 6.
48   Boyer, *By the Bomb's Early Light*, p. 6.
49   Boyer, *By the Bomb's Early Light*, p. 7.
50   Boyer, *By the Bomb's Early Light*, p. 6.
51   Boyer, *By the Bomb's Early Light*, p. 7.
52   Boyer, *By the Bomb's Early Light*, p. 7.
53   Boyer, *By the Bomb's Early Light*, p. 7.
54   Boyer, *By the Bomb's Early Light*, p. 9.
55   Boyer, *By the Bomb's Early Light*, p. 7.
56   John F. Kennedy, *To Turn the Tide* (New York: Harper Brothers, 1962) pp. 208–210.
57   Kennedy, *To Turn the Tide*, p. 211.
58   Abraham Lincoln, "The Perpetuation of Our Political Institutions: Address Before the Young Men's Lyceum of Springfield, Illinois," January 27, 1838, https://www.abrahamlincolnonline.org/lincoln/speeches/lyceum.htm (accessed May 24, 2023).
59   David A. Nichols, *Ike and McCarthy: Dwight Eisenhower's Secret Campaign Against Joseph McCarthy*, (Simon & Schuster: New York, 2017).
60   We did not have these drills in my school, but I've talked to many, many people who were children at the time and remember these "duck and cover"

       drills vividly and angrily.
61   Boyer, *By the Bomb's Early Light*, p. 15.
62   Giulio Douhet, *The Command of the Air*, trans. Dino Ferrari (New York: Coward-McCann, Inc., 1942), pp. 57–58.
63   "We seem to be making just the opposite mistake about air strategy to that we made before the Second World War. Then we grossly over-estimated the capabilities of our bombers to deliver their bombs to their targets, and we grossly under-estimated the amount of chemical bombing [i.e. conventional bombing] which a civilian population could withstand.... Moreover, it is often the same individuals, who before the Second World War espoused Douhet's theories of military decision by air power alone and later found that civilian morale was much higher than expected, who now as a reaction swing to the other extreme." Blackett, *Atomic Weapons and East-West Relations*, p. 29.
64   Bernard Brodie, "Strategic Bombing: What It Can Do," *The Reporter*, 28, no. 552 (August 13, 1950), p. 30.
65   David E. Lilienthal, *Change, Hope, and the Bomb* (Princeton, NJ: Princeton University Press, 1963), pp. 18–19.

4. REFRAMING IT ALL

*It is possible because people don't
hold on to things they don't value.*

We've seen that when people first started trying to figure out what nuclear weapons could and couldn't do, they were operating under two handicaps. First, they were anxious and afraid. The sheer power of nuclear weapons frightened them. And they weren't trying to formulate their ideas about these weapons during a time of quiet contemplation. They were going from crisis to crisis in a confrontation with a powerful adversary. To make matters worse, they were trying to write the manual after only having used the weapons two times.

Given those obstacles, how did they do? Is there some way to assess how accurate those first assessments of nuclear weapons were? Can we gauge how close they came to reality? Let's start with a broad review. Instead of diving deeply into one assumption at random, trying to gauge its accuracy, and then moving on to the next, let's do a quick review of seven or eight of the most important assumptions and see where that leaves us. If they're mostly close to the mark, then probably most beliefs about nuclear weapons from that time were roughly correct. If they're wide of the mark, that would mean that we have reason to dig deeper.

FIRST CUT

One of the first assumptions made about nuclear weapons was that since they rearranged the fundamental constituents of matter—splitting the very atoms that make up the universe—they were clearly revolutionary. After all, Albert Einstein said, "Through the release of atomic energy, our generation has brought into the world the most revolutionary force since prehistoric man's discovery of fire."[66] Whatever the past had been like, people concluded that this new era would surely be radically different.

A corollary to this assumption was that all the old ideas about how the world worked would probably have to be rethought. Humanity was entering a new and entirely uncharted phase of existence. Some people even said that the past would henceforward be divided into three eras rather than two. After the arrival of nuclear weapons, they said, the history of civilization wouldn't just be separated into BC and AD, with a dividing line at the birth of Christ. Instead, there would be BC, AD, and another dividing line between the "Pre-atomic" and the "Atomic" eras.

As far as weapons were concerned, these were obviously the greatest tools of destruction that had ever been invented. Therefore, people concluded that all other weapons were now obsolete. Soon enough, people were saying that since "[t]he atomic bomb is the Absolute Weapon: all conventional armed forces are outmoded."[67]

In military terms, such powerful weapons would certainly change how wars were fought. So, nuclear weapons were quickly labeled the "winning weapon," and people began telling each other that they assured victory in any war.[68] If the United States chose to attack another country, they said, the outcome was now a foregone conclusion.

And these weapons would have a profound impact on defense as well. No nation would ever dare attack a nuclear-armed state—the possibility of nuclear retaliation, it seemed to them, ruled that out. From now on, nuclear-armed countries would be as safe as fortresses.

A few observers drew an even more radical conclusion: nuclear weapons would eventually make all war impossible.[69] Once enough countries had the weapons, no one could risk going to war for fear of the nuclear devastation that would follow.

But people thought the weapons were so awe-inspiring that their impact would reach beyond war. They assumed that nuclear weapons would usher in a radical change in international relations, too. The U.S. secretary of state, for example, was sure that diplomacy would be completely transformed. James Byrnes, impressed by nuclear weapons, became convinced that they "assured ultimate success in negotiations."[70] Any country that had nuclear weapons would be able to dictate terms. Who would risk a nuclear attack by saying "no"?

And in terms of world standing, nuclear weapons were sure to separate the "great" nations from the second-class ones. If you possessed nuclear weapons, you would always get a seat at the highest table. Countries that lacked the technological sophistication or the resources to build these new weapons would be relegated to the seats farthest back in the room.

Finally, nuclear weapons were so valuable and so obviously powerful that many people asserted they would always be with us. How could you ever convince the world to eliminate nuclear weapons? Which national leader would agree to lay down these all-powerful, war-winning weapons? They were such extraordinarily valuable weapons of war that the only way they would ever go away is if war itself were abolished.

These nine assertions are a remarkable collection of claims. If the emotions people felt in those first days reflected reality, then all of these claims will likely turn out to have been true. If the emotions were overblown, somehow, then we can expect that this first set of assumptions about how nuclear weapons would change the world will turn out to be false. So, how did these first estimations and assumptions turn out?

Not very well.

Byrnes's optimistic claim that nuclear weapons gave the United States a free hand in diplomacy was the first to fall. When he flew to Europe to negotiate with Soviet representatives over who would have influence over which parts of post–World War II Europe, he came back chastened and complaining. The Soviets, he said, were "stubborn, obstinate, and they don't scare."[71] And despite repeated diplomatic threats issued by U.S. diplomats, the Soviet Union continued to dominate Eastern Europe for nearly half a century. Nuclear weapons, it turned out, did not assure success in diplomatic negotiations.

The Korean War, fought with conventional weapons and inflicting more than a million casualties, quickly made a mockery of the idea that nuclear weapons made war impossible. And in the years since, there have been more than one hundred wars of varying intensities in all parts of the globe.[72] People had apparently overestimated nuclear weapons' ability to restrain the urge to fight.

The idea that nuclear weapons had made conventional weapons obsolete was exactly backward. Since World War II, conventional forces have been used again and again. But nuclear weapons? Not even once.

The related claim that nuclear weapons would be decisive in war was also debunked by the Korean War, which ended in a draw despite the presence of the United States and its nuclear weapons. And events since have provided even stronger proof that possessing nuclear weapons doesn't assure victory. The defeat of the United States by Vietnam in 1975 and the defeat of the Soviet Union by the Afghans in 1989 demonstrated that nuclear-armed states not only didn't always win but could lose wars.

IT IS POSSIBLE

The claim that no one would ever dare attack a nuclear-armed country seemed self-evident—it is generally fatal to pull the whiskers of a tiger. Yet it still happened. The Middle East War of 1973 and the war fought by the Argentines with the British over the Falkland Islands in 1982 were both wars where non–nuclear-armed states attacked nuclear-armed ones. The notion that no country would ever dare attack a nuclear-armed one turned out, like so many of those early assumptions, to be a mistake born of exaggeration.

The notion that nuclear weapons are the key to prestige and great power status is belied by the fact that a number of states—Germany, Japan, South Korea, Brazil, South Africa, Mexico, and others—have shown that prestige and influence can be had without nuclear weapons. Although people sometimes argue that nuclear weapons must be the key to world importance because every member of the Security Council in the United Nations has them, the fact is that of the five permanent members of the Security Council, only one was a nuclear-armed state when it joined—the United States. The others—China, France, Russia, and the United Kingdom—all acquired their nuclear weapons after they had seats on the Security Council.

And, finally, even though the proponents of nuclear weapons continue to claim that nuclear weapons are "revolutionary," no serious historian now divides the past into "Atomic" and "Pre-atomic" eras.

This review of early assumptions is not a complete survey. Examining a handful of examples is not a systematic review of the entire field. All it shows is that some people held some ideas that turned out to be wrong. So, we can't draw too many definite conclusions from this quick review. We can say, however, that these mistakes were not small mistakes. The idea that no country will ever dare to attack you is no minor matter when it comes to planning your national security.

This collection of mistakes shows that in those first days after nuclear weapons appeared, with so few facts to go on and with emotions running high, the people trying to understand the importance of nuclear weapons often missed the bullseye. They tended to shoot wide of the mark, sometimes very wide of the mark. And their misses were all in the same direction—in the direction of exaggeration. Nuclear weapons would change everything. Nuclear weapons could assure you always got whatever you wanted in any diplomatic negotiation. Nuclear weapons would make wars impossible. Again and again, people assumed that nuclear weapons' power to influence events was greater than it was.

What this means is not that we can dismiss all the thinking that has been done about nuclear weapons in the past. This brief review doesn't mean that we've shown that early ideas about nuclear weapons were all wrong. What these early mistakes give us is a license to go back and recheck that earlier work.

These mistakes are enough to justify reopening this cold case, so to speak. They give us permission to reinvestigate those initial assumptions and beliefs about nuclear weapons and scrutinize them to see if they are true. And since current policy is still largely based on those first Cold War assumptions, they give us the right to be skeptical of nuclear weapons beliefs in general.

So, let's go back and pull the files out of storage. Let's start at the beginning and reexamine all the important facets of the case with an objective eye. We'll begin with one of the key assumptions. For the rest of this chapter, let's focus on one of the most important conclusions that people leaped to about nuclear weapons: that they would probably always exist.

FOREVER WEAPONS

Ask almost any nuclear weapons expert if they think nuclear weapons will be abolished any time soon and they will almost certainly say that nuclear weapons will only go away when human nature changes and war is abolished. The weapons are too valuable, they will assert, for any serious nation to give them up. It is an idea that has been repeated many times. But if you look more closely, it is a rather remarkable claim. And it turns out that other people in other eras have believed that the weapons they possessed were the best, would always be the best, and therefore would always dominate war.

Throughout the Middle Ages, for example, the heavily armored nobleman who was mounted on a charger was considered the pinnacle of military power. Few were surprised when the French opted to make knights the centerpiece of their national armies. But at Crécy and Agincourt, the French discovered to their dismay that noblemen on horseback could be defeated by British yeoman armed with longbows. Similarly, the Carthaginians committed to elephants as the ultimate weapon and gave them pride of place in their armies, only to be destroyed when it turned out that Rome had figured out how to defeat those elephants. British officials were convinced that battleships were the most important measure of a nation's strength. Their infatuation blinded them to changes in technology, and they poured millions into building these warships long after it should have been plain that they would soon be obsolete. World War II brought the realization—with a shock—that torpedo boats, submarines, and dive bombers had ended the reign of battleships as "must-have" weapons.

IT IS POSSIBLE

Refusing to think about what might happen if a weapon becomes obsolete can lead to military defeat (or worse).

Nuclear weapons, I would argue, bewitch us. In the same way that medieval rulers could only see knights and British admirals could only see battleships, nuclear weapons cast a spell on us so that they fill our whole field of vision: We see only the gigantic monster looming over us, and reality fades out of focus into the background. This is all very much like the way a spellbound person can only see what the spell that has been cast on them allows them to see. Nuclear weapons gaslight us, insisting they are the center of our world, constantly trying to convince us that the dark, devastated future they conjure in our minds is reality. But of course, they're lying.

For the last seventy years, nuclear weapons have twisted our worldview. This distortion is especially present in the idea that nuclear weapons can never be eliminated. Seen from the nuclear perspective, this assertion seems like absolute reality. Everyone wants powerful weapons, and nuclear weapons are the most powerful weapons in the world. They will always be powerful. So, they will always exist. As long as human beings long for weapons of power, nuclear weapons will dwarf humankind.

Because experts have been saying this sort of thing for decades, we have long since accepted that it must be so. Maybe you're reading this and thinking, "Well, that does sound kind of true." But look closer, change your perspective, focus on reality, and suddenly this way of thinking transforms into a strange and oddly unrealistic way of looking at the world.

Let's begin with "lasting forever." At some level, you already know this can't be true. No product of human hands lasts forever. Work your way through the evidence of history and you will find there are no exceptions. Percy Bysshe Shelley made this point vividly in a sonnet he wrote in 1818.

> I met a traveler from an antique land
> Who said: Two vast and trunkless legs of stone
> Stand in the desert. Near them, on the sand,
> Half sunk, a shattered visage lies, whose frown,
> and wrinkled lip, and sneer of cold command,
> Tell that its sculptor well those passions read
> Which yet survive, stamped on these lifeless things,
> The hand that mocked them and heart that fed:
> And on the pedestal these words appear:
> "My name is Ozymandias, King of Kings;
> Look on my Works, ye Mighty, and despair!"
> Nothing beside remains. Round the decay

Of that colossal Wreck, boundless and bare
The lone and level sands stretch far away.

Of course, nuclear weapons will one day go away. Everything built by human hands does. Does anyone out there still have a Blackberry? Are you listening to music on an eight-track tape? Are you looking for a job by scanning the want ads in the newspaper? How could we have been talked into believing that nuclear weapons will be immortal when we know that no other kind of technology ever has been? Somehow, we forget what we know when experts insist—with an authoritative tone of voice—that nuclear weapons will be with us "forever."

But second, and much more importantly, think about the rationale for why this "lasting forever" will occur. Nuclear weapons advocates regularly and repeatedly say that it's "because they can't be disinvented." Here, for example, is NATO's Deputy Assistant Secretary General for Weapons of Mass Destruction Policy, Guy Roberts: "Unfortunately, the weapons we've invented cannot be uninvented. We must live with them.... Living with destructive technologies is our lot, the modest punishment we must bear for progress. The bomb is with us to stay. It is, after all, the ultimate guardian of our safety."[73]

I spent years struggling with this one. "How can we disinvent nuclear weapons?" I would ask myself, pacing up and down, slamming my fist into my palm in frustration. "Maybe we could put the formula in a special vault at the United Nations with five separate locks and then limit the access to just a handful of trusted...." It took me, I am embarrassed to say, years to figure out what was going on here—why making progress in arguing against this question was so hard. But once I understood, it reminded me (again) why the pro-nuclear people have won so many of the debates and why the record of seventy years of failure by anti-nuclear forces says nothing about the merits of the question.

The trick here, again, is framing. If you accept their framing, you'll never get out of the box. What I finally did (and what we all have to do) is challenge the hidden premises behind the questions. If you let your opponent define the game, you always lose. "The rules of this game are we flip the coin, and if it comes up heads, then I win; and if it comes up tails, then we flip again." This is a contest you can't win. Flip the coin as many times as you want; it won't change the outcome. Only challenging the rules of the game gives you a chance. Once I finally understood that the issue had to be reframed, a whole new world opened up. Nowhere is this more clearly demonstrated than with the claim that nuclear weapons will always exist.

Nuclear weapons advocates assure us that there's no point in even thinking about eliminating nuclear weapons. After all, they say, you can't disinvent the weapons. And this "you can't disinvent them" objection, when you first hear it, sounds plausible. But if you stop and think about it, it's a trap. Because, after all, disinvention is a process that doesn't exist. Name one piece of technology that you know of that has been disinvented.

One way to get to the bottom of something you can't understand is to concretize it—make it real. So, if you're not sure about disinvention, make it real. Ask yourself, what would this process of disinvention look like? Imagine a large room with multiple workbenches. At each one sits a person in a lab coat surrounded by equipment of one kind or another. A guy comes in (he's dressed in a suit and looks like he might be management) carrying an IBM PC from the 1990s—metal case, two floppy drives, and a whopping 384K of memory. He puts the heavy box down on a workbench and says to the person sitting there, "This one's a rush, Victoria. The boss wants it disinvented by Thursday. If you need anything or if you want Alex to help, let me know." Has a room like this ever existed anywhere? Have any conversations like this ever occurred? Of course not.

And that's how they win. Here's what nuclear weapons advocates have asked you to do. They've asked you to figure out how to eliminate nuclear weapons, and the only tool they've given you to do it with—disinvention—is a process that is entirely imaginary. They're saying, in effect, "Here, son. Dig a hole with this imaginary shovel."

The argument seems plausible because it's absolutely true: you can't disinvent nuclear weapons. But that's not because they're amazing weapons. It's because disinvention doesn't exist. Saying nuclear weapons will always exist because they can't be disinvented is like saying I will always be alive because I can't be reverse born. It's true that I can't be reverse born, but sadly that doesn't mean I'll live forever. It turns out there is another way for me to stop living: We call it dying. And it turns out there is also another way for nuclear weapons to go away—one that the advocates for nuclear weapons, apparently, don't want you to think about.

The key to the entire endeavor, the key to eliminating nuclear weapons is to challenge the thinking of the nuclear weapons policy elite. They have cast the problem in a particular way, but it seems obvious, common sense repeatedly tells us, that there are serious problems with the way they have imagined the field, structured the thinking, and asked the questions. This is not a new idea. As early as the 1980s, some influential observers were asserting that there were serious problems with the approach that the policymaking elites were taking to this critical issue.[74]

It is, in fact, entirely possible for nuclear weapons to go away. To see how all you have to do is first set aside the framing that nuclear weapons advocates have put on the problem—"disinvention"—and see the problem in a different way. We need to take off the "disinvention" goggles we've been wearing for so long, toss them aside (you won't be needing those anymore), and hold a new lens up to our eyes. To understand how nuclear weapons might go away, examine them using the lens we use for any other kind of technology: look at them, in other words, through the lens of technological evolution.

Our tools—originally limited to simple stone knives and scrappers but eventually including complicated, manufactured pieces of technology like smartphones—have been evolving for as long as humans have used them. Human beings always try to improve their tools. And the process that controls that evolution has been relatively constant across the millennia. The evolution of technology moves through a cycle with four phases: invention, adoption, use, and eventual abandonment. Looking at the nuclear weapons question in terms of technological evolution yields an entirely different result than trying to see the problem through the lens of "disinvention." By reframing the issue, we can change the problem completely. And this change of perspective gives us a pathway to elimination that we hadn't ever noticed. Technology doesn't go away by being disinvented; it goes away when it gets abandoned.

So, if we want to understand how nuclear weapons might go away, we should study abandonment. Instead of staring vainly at "disinvention" with a bewildered look on our faces, we need to go back through the history of various technologies that have been abandoned, take the process of abandonment apart, and figure out what makes it tick.

Unlike disinvention, abandonment is a process we're all familiar with. Who hasn't set aside a pile of things they don't use anymore just inside the door to the garage, waiting for some Saturday when there'll be time take them to the dump? The criteria they used to sort their possessions into the "to keep" and the "to toss" piles are roughly the same criteria that humanity uses on all the different kinds of technology we have adopted, used, and eventually abandoned. And it is the same criteria that people will use with nuclear weapons when they finally get around to confronting them as they actually are.

Advocates of nuclear weapons talk about the invention of nuclear weapons as if it were some sort of "once you've done it, you can never go back" process. They act as if now that the weapons have been invented, they will always be around. But at least six thousand years of technological innovation tells us that invention isn't the crucial step. Adoption is what matters. Widespread adoption and use are what make a piece of technology important. And adoption is an entirely reversible process. Anything that can be adopted can also be abandoned.

In some ways, it is surprising that some entrepreneur or venture capitalist hasn't written a scathing article ridiculing the "disinvention" argument that has so dominated nuclear weapons debates for decades because people who invest in technology are deeply familiar with the actual ebb and flow of technological evolution. Their financial well-being depends on knowing the difference between invention and adoption. They know that a new piece of technology can be amazing; it can be the latest gee-whiz breakthrough stuff. But if it doesn't get adopted, if it isn't widely used, it will never matter.

So, how does this adoption, use, abandonment cycle work? Again, examining actual cases brings the issue into focus. Which technologies get adopted and why can be clearly seen in historical examples. For example, sometime after 221 BCE, the shipbuilders working for Ptolemy IV, then the ruler of Egypt, were asked to build the largest warship that had ever been built. Early warships were powered by a row of oarsmen on each side. At some point, a second deck was added above the first and—voilà!—twice as much power. More decks were added, more rowers per oar, and eventually the proportions of these ships became truly massive. Some, apparently, were so large they could carry around large catapults on their decks.

In response to his command, Ptolemy's engineers built a giant, almost unbelievable ship with four thousand rowers. Everything about this ship was oversized. Most ships of that era carried a complement of soldiers that could leap onto enemy ships and take them over. Ptolemy's monster ship carried 2,850 soldiers—and probably a catapult or two.

The number of rowers and soldiers was recorded, so we can be confident that they are factually accurate. But the exact design of this giant ship—its configuration—is still something of a mystery. There is speculation that it involved connecting two hulls together, each hull with multiple levels of rowers and each oar having multiple rowers working it, forming a sort of giant catamaran. Imagine how intimidating this vessel must have looked. Rows and rows of oars all moving together, almost three thousand warriors standing poised to leap onto opposing ships, and catapults hurling giant rocks. As a symbol of dynastic power, it must have been extraordinarily effective.

So, what happened to this amazing advance in the technology of warfare? It must still be with us, right? After all, once something gets invented, it can't be disinvented, can it? What happened is that Ptolemy's massive ship was used a couple of times and then broken up for firewood.

After a few times in battle, the giant ship design was abandoned. And no other rulers copied Ptolemy's remarkable design—it was never widely adopted. Why? Ptolemy abandoned it and others didn't adopt it because the giant ship wasn't very effective. It turned out that smaller ships were faster, they could dart away from cumbersome ships like Ptolemy's, and, despite its impressiveness, a giant ship with thousands of rowers was virtually useless in battle. As a result, giant ships had all but disappeared by the middle of the third century BCE.[75] The crucial criterion for both adoption and abandonment was utility.

Or consider another example from World War I. The Paris Gun was a giant cannon built by the Krupp Works for the German army. It was a truly remarkable weapon that relied on a collection of cutting-edge advances in military technology. It weighed 256 tons and had an immense barrel that was 112 feet long. The barrel was so long it had to have an additional set of supports to keep it straight, like a suspension bridge. The gun itself was so heavy that if it had been transported on ordinary roads, they would have buckled. It fired a 234-pound shell eighty-one miles, and at the height of its flight, the shell was more than twenty-six miles above the earth's surface, thus making the Paris Gun's shells some of the first man-made objects to reach the stratosphere.

It was, all in all, a remarkable weapon at the forefront of artillery technology. Its size was impressive. And its arrival in the war was frightening. Even though Paris was far behind the front lines, suddenly huge shells began falling on the city. They came without warning. Standing in a street, sitting in a café, returning from the market, suddenly there would be an explosion and dirt and debris would rain down. From March until August of 1918 shells fell out of the sky, killing 240 people and wounding six hundred. You would expect that such a remarkable weapon would quickly change the face of war.

But there are no Paris Guns in the arsenals of the world's militaries today. "Superguns" like the Paris Gun are not sought after by lesser countries hoping to gain prestige and elbow their way into the front rank of the world's nations. The Paris Gun did not play a crucial role in the outcome of World War I. It did not terrify the people of Paris or drive them to beg their government to surrender. Even though building such weapons is still possible, you cannot find Paris Guns on any continent today. They have been effectively eliminated.

"But that's no mystery," you might say, "because they weren't very effective." And you would be right; they weren't very effective. The Paris Gun was so inaccurate, it couldn't hit anything smaller than a city. It also didn't actually kill that many people or do that much damage. It couldn't be fired very often—the wear on the barrel was so great that the entire barrel was removed and replaced after every sixty-five rounds. It required enormous expense and effort. Simply deploying it in the field meant building railroad tracks to the place where it would be stationed. And the hope that it would break the will of the French people by frightening them did not come to pass. Even though the weapon was bigger than any cannon ever previously built, even though mastering the technical challenges involved in hurling a projectile into the stratosphere made the Germans feel patriotic, the weapon didn't survive. In the end, the Germans made their decision based on the practical utility of the weapon. They looked at the results and concluded they should spend their military funds on other, more effective weapons. Again, utility was the criterion that decided the Paris Gun's fate.

So, the answer to the question "How will nuclear weapons go away?" has nothing to do with disinvention. It is not about how moral human beings are or whether we can change human nature. The elimination of nuclear weapons will be controlled by the same criteria that have controlled the decision to abandon any other kind of technology. That is a realist's approach to this problem.

A NEW PERSPECTIVE

In the grand cycle of technological evolution, the driving criterion is quite simple: utility. Tools and technology get adopted, used, or abandoned depending, for the most part, on the answer to one question, "Is it useful?" The ship Ptolemy built was expensive and impressive. It must have inspired awe in the people who saw it. No doubt, it was seen as a symbol of Ptolemy's greatness. But because it wasn't effective in battle—because as technology, it turned out not to be useful—all the awe and symbolism in the world couldn't keep it around. The giant warship was quietly dismantled, and no other kings raced to build similar ones.

The Paris Gun was awesomely big, too. It employed the latest technology. But it, too, found its way into the scrap heap of history because it didn't help to win the war.

And this brings us to the analytical heart of our investigation: utility. This is the criterion we will use to evaluate nuclear weapons. This is what will determine whether they are eventually abandoned or not. Are nuclear weapons useful on the battlefield? Can they be used effectively against an enemy's homeland? And can nuclear weapons be used as threats over the long run? Will keeping large numbers of weapons on constant alert, poised to attack at a moment's notice, eventually lead to disaster? These are the questions that make up a pragmatist's approach to eliminating nuclear weapons. If it turns out that nuclear weapons are useful, then they have to stay. But if it turns out that they are not very useful and are also dangerous, then they have to go.

And that makes sense. If I tried to sell you a car that couldn't go any faster than fifteen miles per hour and the steering would occasionally fail without warning—sometimes turning you into oncoming traffic, sometimes off the road over a cliff—would you buy that car? If I gave you a nail gun that didn't drive the nails in quite far enough, so you had to finish driving them in by hand, and the gun sometimes just randomly fired a nail—even though you hadn't pulled the trigger—into your foot or your assistant or the homeowner, would you use that nail gun? If I tried to send you out on patrol armed with hand grenades that only exploded about fifty percent of the time, and when they did explode about one time in five, they exploded in your hand when you pulled the pin, not after you threw them, would you keep those grenades? No, of course not. You'd never buy that car, use that nail gun, or keep those grenades—because nobody keeps technology that is both virtually useless and very dangerous. Nobody.

So, this is the plan for the rest of the book: we're going to investigate the utility (and danger, which affects utility) of nuclear weapons. And since utility is measured against the real world, the watchword for our investigation will be reality.[76] We know that nuclear weapons are immoral. No one doubts that. Even the people who promote the weapons' value quietly admit that their use would probably be immoral. So, we don't need to argue over that. But we've seen that immorality is not enough to get the weapons eliminated. More than seventy years of failure proves the point. So, let's accept that the weapons are immoral and put that part of the debate aside for the moment. The questions we want to examine are on the other side of the coin. We want to dig into the sometimes-harsh world of reality. A single nuclear weapon could kill hundreds of thousands of civilians in an afternoon. But horrible and immoral as that is, does that really help to win wars? Nuclear war is so fearful that it seems obvious that no one would ever fight such a war. But are we really looking in the darkest places of the human soul when we reassure ourselves that "no one would ever be crazy enough to start a nuclear war"? These are the difficult issues we need to confront. Uncomfortable or not, we want to know the answers. We want to know what we're faced with.

A great deal has been said about nuclear weapons. They have spawned what one scholar called "a truly massive" literature.[77] Many people who oppose nuclear weapons have talked and written about the feelings that nuclear weapons inspire—our awe, our sense that the weapons are immoral, and our horror at the possible destruction. The world has debated these issues for more than seventy years, in fact. Some of the talk has been sensible. Some of it has been biased. Some of it has been ill-informed—based on impressions more than careful thought. We have talked this wretched subject over for so long, is it any wonder so many people want to put the whole uncomfortable, fearful issue out of their minds?

But there is a reason why we've been stuck for seventy years. The deck was stacked; the fix was in. We were pointed in the wrong direction, toward mirages like "disinvention." Now there is an entirely new conversation to be had—a new angle to approach the problem. And this new, cleaner, more realistic approach actually has some promise to settle the matter, to bring us to a resolution. If nuclear weapons are the greatest danger that faces civilization, the greatest potential source of sudden catastrophe we confront, then this journey of exploration and discovery is one of the most important tasks we can undertake. It is the only realistic path to discovering if these weapons can once and for all be put aside—abandoned. I believe that if we undertake that journey with a realist frame of mind, remarkable progress can be made.

So let us take heart and remember what Henry David Thoreau wrote about reality in Walden:

> Let us settle ourselves, and work and wedge our feet downward through the mud and slush of opinion, and prejudice, and tradition, and delusion, and appearance ... till we come to a hard bottom and rocks in place, which we can call reality, and say, This is, and no mistake.... Be it life or death, we crave only reality. If we are really dying, let us hear the rattle in our throats and feel cold in the extremities; if we are alive, let us go about our business.[78]

Like Thoreau, we want reality—not because we are prejudiced against daydreaming, not because we want to live in a cold, hard world with no wonder, beauty, or romance, but because reality is key to survival. It is the map that can lead us out of this ugly and frightening nuclear landscape.

The entire approach to discussing whether we should keep or abandon nuclear weapons has been miscast. Framed as a question of disinvention, it turns out that this way of thinking about the problem ensures that elimination will be seen as impossible. Looking at nuclear weapons through the lens of utility, an entirely different conclusion is possible. When major structural errors emerge after a building has already been erected, even if it has stood for seventy years, prudence demands that the entire building be reexamined—foundation, materials, framework, and every other part. Similarly, it is now clear that it is no more than simple prudence to closely reassess the thoughts and beliefs that guide nuclear weapons policy.

66 "Project on Government Secrecy: Einstein fund raising letter of January 22, 1947," *Federation of American Scientists,* https://sgp.fas.org/eprint/einstein.html (accessed October 27, 2022).
67 Blackett, *Atomic Weapons and East-West Relations,* pp. 70–71.
68 Assistant Press Secretary Eben Ayers recorded in his diary that at a Cabinet meeting in 1946 Truman said that he didn't believe there were more than six nuclear weapons then in the United States arsenal, but he was quick to reassure Ayers that "that was enough to win a war." Gregg Herken, *The Winning Weapon: The Atomic Bomb in the Cold War, 1945–1950* (New York: Vintage, 1982), p. 197. See also the description of a memo prepared by Secretary of War Stimson for President Truman that warned that with nuclear weapons, a small and unscrupulous nation might be able to conquer much larger ones. Herken, *The Winning Weapon,* p. 15. According to Gordon Dean, who was head of the Atomic Energy Commission from 1950 to 1953, he said that the United States found itself in possession of a weapon "almost universally credited with the power to destroy cities on a ratio of one bomb per city, and to end wars on a ratio of two bombs per war." P. M. S. Blackett, *Studies of War: Nuclear and Conventional* (New York: Hill and Wang, 1962), p. 18. "It is well to remember that Britain in 1946 fully and emphatically supported the thesis that a great power could be quickly and cheaply defeated by atomic bombs alone." Blackett, *Atomic Weapons and East-West Relations,* p. 91.
69 "If it is in fact true, as most current opinion holds, that strategic air power has abolished global war, then an urgent problem for the West is to assess how little effort must be put into it to keep global war abolished." Blackett, *Atomic Weapons and East-West Relations,* p. 32.
70 Herken, *The Winning Weapon,* p. 45.
71 Herken, *The Winning Weapon,* p. 53.
72 The number varies depending on your definition of what constitutes a war.
73 Guy Roberts, "The Continuing Relevance of NATO's Nuclear Deterrence Strategy in an Uncertain World," in *Peace and Disarmament: A World without Nuclear Weapons?* ed. Hannes Swoboda and Jan Marinus Wiersma (Brussels: PSE, 2009), p. 65, which is cited in Tom Sauer and Bob van der Zwaan, "U.S. Tactical Nuclear Weapons in Europe after NATO's Lisbon Summit: Why Their Withdrawal Is Desirable and Feasible," *Belfer Center Discussion Paper,* Harvard University, No. 2011-05, (May 2011), p. 33.
74 Fred Kaplan's *Wizard of Armageddon* is essentially a critique of the entire field and their abject failure to create reasonable policies for managing nuclear weapons. The book ends with these words: "The nuclear strategists had come to impose order—but in the end, chaos still prevailed." Fred Kaplan, *The Wizards of Armageddon,* (New York: Simon and Schuster, 1983), p. 391.
75 Lionel Casson, *Ships and Seamanship in the Ancient World* (Princeton, NJ: Princeton University Press, 1971), pp. 97–115.
76 Not the sort of Realism taught in schools of international relations. I like John Mearsheimer. He invited me to speak at Chicago University one time, but I don't hold with his sort brand of Realism. I mean realist with a small "r," everyday realism, the kind that ordinary people have. I mean realism that

values facts and experience above all else, that is suspicious of theories until they are really proved. When I talk in this book about realism, don't think of international relations; think of the way people face hard truths when they're up against it.

[77] John Mueller, *Nuclear Alarmism* (Oxford: Oxford University Press, 2010), p. 63.
[78] Henry David Thoreau, *Walden: or, Life in the Woods*, https://www.gutenberg.org/files/205/205-h/205-h.htm (accessed May 24, 2023).

5. BIGNESS—IT'S OVERRATED

It is possible because bigger is not always better.

So now, with all the preliminaries finally out of the way, let us turn to the practical nature of nuclear weapons. For the next three chapters, let's explore whether nuclear weapons are militarily useful or not. Let's begin with bigness. Nuclear weapons are assumed to be decisive weapons for two reasons: first, their enormous size, and second, their ability to flatten and destroy whole cities. In this chapter, we're going to explore the issue of size and try to understand what it means for a weapon to be big.

This question of size is interesting because if you scratch the surface of the question, you discover an unexpected mystery. We can all agree that, at first glance, it appears that nuclear weapons' bigness is decisive. The reason they made such an impression on people was that the explosions were so large, the flash of light so bright, the roar so loud, and the column of smoke that rose from the explosion so tall. It is not saying anything controversial to say that their influence on world affairs and in our thinking about security comes from their size. Gerard DeGroot, emeritus professor at St. Andrews, Scotland, talks eloquently about this in his book, *The Bomb: A Life*:

> Nothing that man has made is bigger than the Bomb.... The sheer size of the Bomb ... has dominated the minds of those condemned to live in the atomic age, exerting its influence not just on politics, but also throughout popular culture.... Real control over the Bomb will come when something bigger takes its place. H.G. Wells thought that the solution would come in the form of world government. Many atomic physicists clung

to that hope as a solution to the torment of responsibility. A clutch of religious lunatics have promised that faith in their particular Messiah will provide an effective nuclear shield. Ronald Reagan thought the answer might lie in an assortment of satellites and powerful rays.... But whether the answer lies in politics, science or faith, the only thing that seems certain is that the solution to the Bomb's bigness seems far distant.[79]

DeGroot exactly captures the most important source of nuclear weapons' hold on our imaginations: size. We stand in awe before nuclear weapons; they seem to loom high over us. Crane your head back to look up at the towering mushroom cloud—or even just imagine seeing one—and you are humbled and filled with awe.

But this is where the mystery comes in. If you examine the size of nuclear weapons objectively—rather than feeling awe—what you find is confusing and counterintuitive: their size is shrinking. DeGroot is sure that nuclear weapons are bound to grow, and that their largest size has yet to be reached. But the trend seems to be the other way, which is counterintuitive. But it is also the case. Examine the history of nuclear weapons, and you will notice that, over time, the size of the explosions that the bombs are designed to make—their yield—has been steadily decreasing. Think about that. The device that is important because of its bigness has been steadily shrinking. This is like going to the garden shed and discovering that in the ten years since you last used it, your sledgehammer has shrunk to the size of a child's play hammer. How can you knock down walls with that?[80]

The average yield of a strategic nuclear weapon in the U.S. arsenal in the early 1960s was about a megaton (the equivalent of one million tons of TNT). Today the average yield of a weapon in the U.S. arsenal is about 175 kilotons (the equivalent of 175,000 tons of TNT)—roughly eighty percent less than the average yield in the 1960s. In 2022, the U.S. government announced plans to retire the biggest bomb still in its arsenal, the B83.[81] But even with the B83 included in the average, U.S. bombs are five times smaller than they were sixty years ago. There has also been a decrease in the size of Russian nuclear weapons. How can this be? How can the yield of the bombs be getting smaller? Isn't bigness what makes nuclear weapons matter?

Obviously, this slow shrinkage over the years is not accidental. Military planners have apparently been requesting, the U.S. government has been approving, and nuclear warhead construction facilities have been building nuclear warheads with lower and lower yields. The question is: why? Is there some sort of secret, treasonous cabal deep within the heart of the U.S. government and military? Is there some group bent on leaving the United

States helpless in the face of our adversaries? Or, if military planners think that smaller weapons would be better for some reason, what's the reason? Don't they know what makes nuclear weapons important?

If bigness is what confers power, then the size of nuclear weapons should be increasing. Each new weapon should have a larger and larger yield. This is especially so because there are few practical barriers to increasing the size of hydrogen bombs. In theory, you can build a hydrogen bomb as big as you want.

A brief technical aside: hydrogen bombs consist of a fission bomb (like the one dropped on Nagasaki) that acts as a detonator, and a physically separate secondary bomb containing the hydrogen fuel. When the fission bomb goes off, its atoms split, releasing an enormous surge of X-rays and neutrons. The X-rays compress and heat the hydrogen fuel to hundreds of millions of degrees, pressing atoms together forcefully enough to cause them to fuse. Fusion releases energy, which sets off more hydrogen, much like adding wood to a fire. If you want to make a bigger hydrogen bomb, all you have to do is keep adding stages of hydrogen isotopes. There is no theoretical limit to the size of such a bomb.

Think about that. Once the H-bomb was invented in the early 1950s, nuclear weapons could be made as large as you wanted. It was practical and feasible to keep building bigger and bigger, more and more impressive nuclear weapons. The Soviets once exploded a weapon that was estimated to have had a yield of fifty-six megatons (the equivalent of fifty-six million tons of TNT). So, if you can make them bigger, and bigness is what makes nuclear weapons important, why wouldn't you just keep making them bigger? Why would you slowly make them smaller over time?

BIGNESS QUOTIENT

How can we explain this strange contradiction? Could there be some secret involved here? Or is there something about bigness that we've failed to grasp? Is it possible that bigger bombs aren't always better bombs?

Perhaps the best way to explore the "bigger is always better" notion is to begin with a hypothetical. Imagine that you could boil everything about a weapon down to one number, and that number would measure how effective that weapon was. Call it the weapon's "Bigness Quotient," or BQ for short. Using the BQ, all weapons can be measured on the same scale and compared to one another. So, a pistol, for example, might have a relatively low rating, say, 120 BQ. A howitzer, we could imagine, would have a higher rating—say, a thousand BQ. And a nuclear weapon would have the highest BQ possible—say, a million BQ.

IT IS POSSIBLE

Conceived this way, solving a military problem simply becomes a question of how much force to apply. Start by trying to solve the problem with a weapon with a relatively low BQ. If that doesn't work, try the weapon with the next highest BQ. And so on. First, try a knife. If a knife doesn't work, try a pistol. If a pistol doesn't work, try a rifle. If a rifle doesn't work, try a cannon. If a cannon doesn't work, try a nuclear weapon. Weapons are, according to this conception, basically all the same. The only question is how much BQ they have.

This way of thinking about military power reminds me of a classic scene from certain old science fiction movies: Our hero is strapped to a table with wires and strange devices attached to him. The mad scientist shouts at him, demanding that our hero tell him where the gold is hidden, what the secret formula is, or whatever. The hero bravely refuses, struggling against his restraints. The mad scientist reaches for a large dial and turns it slowly. We watch a close-up of the dial in horror as more and more voltage is applied. Our hero twists and convulses in agony, and we wonder silently to ourselves, "How much can he take?"

From the mad scientist's point of view, the only question is how much electricity will be necessary to force our hero to give in. He doesn't ask himself, "Will I have to attach more or different devices?" He doesn't wonder whether the wires are attached to the right locations on our hero's body. He simply wonders, "How much more?" This is exactly the way some people view military problems. In their minds, military problems are all the same: in every situation, the only question is how much military force to apply.

But this way of thinking about weapons can't be right because there are obvious examples that contradict it. If, for example, you were fighting underwater, you'd be much better off having a speargun than a howitzer, because field artillery doesn't work very well under water.[82] If you needed to kill silently (so you won't be caught), a knife is a much better weapon than a pistol. Clearly, bigness doesn't capture all the subtleties of the issue. In fact, thinking just in terms of bigness can lead to catastrophic mistakes.

For example, imagine that we are the board of directors of a community bank in the 1950s. There have been some robberies in the area lately, and we've decided to hire a guard to stand in the lobby. As soon as the board has voted for a guard, someone asks, "What sort of weapon should we get for him?" (It's the 1950s, so we're probably all male directors and it is unsurprising that we're assuming the guard will be a man.) Somebody else says, "I don't care what weapon we get for him, as long as it's the most powerful weapon available. Go big or go home, I say." Around the table, heads nod in agreement.

Now picture the scene. The lobby of the bank is full of customers, waiting in line to speak to a teller, filling out their deposit slips, or waiting to talk to the manager. The guard stands in his uniform, slightly bored, rocking back and forth on the balls of his feet, with three sticks of dynamite tucked into his big leather belt. A robber comes in, steps up to Jenny at the third window, and says, "Gimme all the dough." What is our guard going to do?

If he lights a stick of dynamite and throws it at the robber, he'll kill the robber and Jenny and most of the customers in the lobby—possibly even himself. If he doesn't throw the dynamite, the robber will get clean away. He might be able to threaten the robber and make him think he'll throw the dynamite, but once he lights the fuse, there's only so much time to negotiate. And there's no guarantee that threatening will work. Too late, he realizes that, although he is armed with the most powerful weapon available, the one that makes the biggest explosion, the weapon he's been given doesn't really help him do his job. Later, when the board meets in emergency sessions, they realize ruefully that there may be other criteria that define a weapon's usefulness besides bigness.

WHAT WEAPONS ARE

So, let's take a step back and ask ourselves, what are weapons? Are they all the same—just variations on the same fundamental thing? Or are there important characteristics that differentiate weapons? Weapons, when you stop and think about it, are tools—a subset of the larger class. Tools are implements that extend the natural abilities of our bodies. Our fingernails are only so strong, so we extend their ability to dig and claw with a long knife. Our fists can bruise and the knuckles get skinned, so we invent a hammer for hitting. Tools increase our ability to manipulate the world around us. It's usually hard to loosen a firmly tightened nut with your fingers. A wrench makes it much easier.

One of the key qualities of tools is that they are situational. In other words, their usefulness depends on the circumstances. A flathead screwdriver is the right tool for taking the cover plate off an electrical outlet on your wall. It would be a terrible choice if you were going to use it to paint the side of a barn. A sieve is great for separating small stones from soil. It would make a lousy bucket. A sledgehammer is the best tool for breaking up an old sidewalk; use it to prune trees, and you risk getting injured as well as doing a poor job of pruning. The value of a tool is not based on some imaginary uniform measure of its bigness. Its value is determined by how well its characteristics match the task at hand.

Weapons are like tools in this regard. They can be highly effective in one set of circumstances but useless in another. A grenade isn't ideal when you and your enemy are both in a small elevator, a battleship would be wasted in a zone where enemy aircraft can bomb it from above, and land mines buried on the ocean floor probably won't stop scuba divers (unless they happen to walk on the seabed). Judging the effectiveness of a weapon involves predicting the circumstances where you'll need it and then assessing whether the characteristics of that weapon match the situation. Are you going to kill someone stealthily? A knife could be your best choice. But if you are planning to provide covering fire from a distance, a knife won't work.

Big weapons can be useful—when the situation calls for big weapons. But sometimes the situation calls for something other than bigness. Sometimes it calls for quietness, for waterproofness, for discreteness, or for a myriad of other characteristics. In fact, the sheer number of different types of weapons is proof that there are many, many different circumstances in which we want to use weapons.

So, the question we need to ask when trying to evaluate nuclear weapons is not "how big are they?" That tells us almost nothing about their effectiveness. What we need to ask is this: "In what circumstances are they likely to be used? Do their characteristics match that situation? And does that situation arise very often?"

As a first step in thinking about how weapons are used, let's think for a minute about targets in war. This will help us judge when we can best use nuclear weapons. The most frequent target in war (I would argue) is a soldier. The object of war, after all, is to defeat your adversary's military; and a country's military is composed of soldiers. After soldiers, the next most frequent target is vehicles that carry soldiers or weapons. Tanks, trucks, ships, helicopters—these are all things you generally need to destroy to win. The third category of target is structures: barracks, headquarters, intelligence buildings, radar arrays, and so on. Finally, there are some relatively rare targets that are larger than a building—like airfields or supply depots. Given that the three most common targets are soldiers, vehicles, and buildings, and that other size targets are relatively rare, I would argue that the maximum useful size for an explosive weapon is one big enough to destroy a building. In the relatively rare cases where you want to blow up a target that is larger than a building—like an airfield—you can simply use more bombs that create building-sized explosions.

What's wrong with blowing up more than just the thing you're targeting? Why not kill one hundred soldiers at once? Does it make a difference if you leave behind a building-sized crater or a neighborhood-sized crater? Why not go as big as you can?

One practical problem is that often, on the battlefield, your guys are pretty close to their guys. This makes it problematic to use a really big weapon. A direct hit on a frontline unit is likely to kill a significant number of your own soldiers. When you deliberately kill your own soldiers, it tends to put a damper on morale.

You could, of course, use nuclear weapons farther back from the front line. But if winning wars is about defeating soldiers, why would you waste your energy blowing up things far behind the lines? This is like attacking the tail of the snake, rather than trying to cut off its head. And larger explosions will blow up things that you might want to keep intact, like bridges, hospitals, factories, or other valuable resources.

Size, it turns out, can limit the utility of explosive weapons once the explosion gets beyond a certain size. Really big weapons are often more clumsy than effective.

THE HISTORY OF BIGNESS

The history of war confirms that big weapons often turn out to be surprisingly ineffective. Dreadnoughts were, after all, the biggest battleships of their day. But it turned out they were not the most effective weapons. Smaller weapons—torpedo boats, submarines, and dive bombers—were all able to defeat these great iron warships.

And dreadnoughts are not simply a one-off exception. In the ancient world, elephants were sometimes considered unbeatable weapons. The Carthaginians, for example, relied on elephants. These huge creatures towered over all soldiers and were almost impossible for a single soldier to kill. They could use their hideous strength to trample and toss men aside like loaves of bread. Against untrained troops, who would often break and run at the mere sight of war elephants, they were highly effective. Viewed from a distance, the assumption that elephants would be decisive on any battlefield looks sound.

But dig into the details, look more closely and it turns out that elephants had other, negative characteristics that outweighed the benefits that came with their size. They were slow and required enormous amounts of food and special handling. They did not survive well in cold climates. During Hannibal's famous crossing of the Alps, for example, almost all his elephants died from the harsh conditions and cold. And elephants can be beaten. Well-trained troops can defeat elephants by working together.

But these drawbacks are not the most serious problem. The greatest difficulty was that elephants were far stronger than the men who rode on their backs and were supposed to control them, which meant that if the elephants became excited, they would do whatever they felt moved to do. Sometimes in battle—in fact, quite often in battle—elephants ran amok, at which point they were as great a danger to their own troops as the enemy's. At the battle of Zama, for example, the decisive battle that led to the Carthaginians' defeat in the Second Punic War, the Carthaginian elephants inflicted little damage on Rome's soldiers. But when several ran mad through the left wing of the Carthaginian forces, they trampled their own soldiers and threw that part of the army into disarray. The money and training the Carthaginians spent on elephants did not save Carthage and, at the battle of Zama at least, may have played a decisive role in their losing the war.

Elephants, dreadnoughts, and other rarely remembered large weapons are reminders that size is not necessarily proof of effectiveness. The biggest is not always the best. If you had to fly to Europe, for example, would you choose to go by dirigible? Dirigibles remain the biggest flying ships ever. The Hindenburg was 804 feet long—three times longer than a 747 and only seventy-nine feet shorter than the Titanic. Dirigibles are big, in other words. But flying a dirigible would be slow and potentially more vulnerable to storms than modern commercial aircraft. When you fly, the most important criterion is not the size of your aircraft. The same can apply to weapons.

One more example, this time an imaginary one. Imagine that you were the leader of Costa Rica and your scientists had come up with a way to create hurricanes. You would congratulate them on their amazing achievement and think happily that you have vastly increased your security. (Since Costa Rica renounced having a military in 1949, you would also be going against your country's long tradition of peacefulness. But it's an imaginary example.) Think about the power of a man-made hurricane. It is a weapon that would almost surely overawe and intimidate your adversaries. But there are some practical difficulties.

Say that you're upset with Nicaragua. Say there's a border dispute between your two countries. (There is, in fact, a long-running disagreement over borders between these two countries.) Can you use this new mega-weapon against Nicaragua? Of course you can't, because any hurricane that hits Nicaragua will also do damage to Costa Rica, too. Rain, ferocious winds, flooding, and all the other effects of a hurricane would inundate some parts of Costa Rica. And, of course, there is always the possibility that the hurricane would get out of control and veer off course, hitting your country square on and missing Nicaragua altogether. A "Hurricane weapon" would be the biggest weapon ever, but it would also be dangerous and difficult to control. There are some weapons that are simply too big to be useful.

BIGNESS—IT'S OVERRATED

Although it's rarely emphasized in the public debate, this notion that nuclear weapons may well be too big to be useful has been a common theme in the private discussions of experts and government officials. Go back through formerly secret meetings, and the idea that nuclear weapons are too big comes up with surprising frequency. For example, when it became clear that it might be possible to build hydrogen bombs—with yields equivalent to millions of tons of TNT—the United States government established the General Advisory Committee and asked it to weigh in on whether the United States should build such mammoth weapons. The committee was chaired by J. Robert Oppenheimer, the scientist who had led the effort to develop atomic bombs, and consisted of distinguished scientists. The report they issued expressed unanimous opposition to building hydrogen bombs.

> It is clear that the use of this weapon would bring about the destruction of innumerable human lives; it is not a weapon that can be used exclusively for the destruction of material installations of military or semi-military purposes. Its use therefore carries much further than the atomic bomb itself the policy of exterminating civilian populations.... By its very nature, [this new weapon] cannot be confined to a military objective but becomes a weapon ... of genocide.[83]

In the judgment of the scientists, industrialists, and engineers who wrote this report, hydrogen bombs would be too big. Because the circle of destruction the bomb created was so large, even if a bomb were dropped exactly on a military installation, it would still devastate a great area beyond that installation. Therefore, except when it was used against the most isolated targets, far from any human habitation, it would inevitably kill large numbers of civilians.

Assuming that bigness determines military effectiveness is a mistake. When the carpenter sends his assistant to get a tool, he doesn't say, "Darren, go out to the truck and bring back the biggest tool for the job." He says, "Darren, bring back the right tool for the job."

If you remember, in that first quote at the start of this essay, Gerard DeGroot said, "Real control over the Bomb will come when something bigger takes its place." This isn't right. The truth is that real control over nuclear weapons will come when something better takes its place.

IT IS POSSIBLE

And that "something better" might well be—in fact, it almost certainly will be—something smaller. The value of a weapon, it turns out, is not determined by its size but, rather, by its effectiveness.

* * *

So, does the size of nuclear weapons matter? Not much. Bigness is simply one characteristic that a weapon can have. Sometimes size is useful, sometimes not. In the case of nuclear weapons, their size is so large that it's a hindrance. Size holds nuclear weapons back from being used in almost every situation.

The bigness of nuclear weapons matters. It has shaped our world—done something crucially important. But the important thing that bigness has done is not to win World War II; the important thing it's done is not to solidify the series of alliances that structure the current world order; and the important thing it's done is not to prevent another world war. The most important thing that the bigness of nuclear weapons has done is to fool us—to fool us into thinking they are more important, more effective, and more useful than they actually are. Yes, the towering cloud is impressive. Yes, explosions can create lots of destruction. Yes, they are certainly dangerous. But in the end, they are not that useful. It's hard to put them to any practical purpose. Ultimately, they are not like Zeus' thunderbolts. They are like a bull in a china shop: big and strong but also clumsy, bumbling, destructive, and virtually useless.

[79] Gerard J. DeGroot, *The Bomb: A Life* (Cambridge: Harvard University Press, 2005), pp. iix-x.
[80] Understanding the size issue is made harder by the fact that it's hard to compare the size of one nuclear weapon to another. It's not like putting two growing boys up against a wall, measuring how tall each is, and then comparing the numbers to see who's tallest. With yield, the relationship between the force of the explosion and the destruction on the ground is not a one-to-one thing—which sounds a little strange, so to see this in concrete terms, let's compare the Hiroshima bomb to a one megaton hydrogen bomb. The bomb dropped on Hiroshima had a yield of sixteen kilotons, which is the equivalent of sixteen thousand tons of TNT. A one megaton bomb is the equivalent to one million tons of TNT. If measuring the size of nuclear weapons was like measure the height of two boys, then you would expect the destructive area of the one megaton bomb would be 66.7 times bigger than the Hiroshima bomb (one million is 66.7 times more than sixteen thousand). But if you draw concentric circles of destruction on a map—massive destruction in the innermost circle, medium destruction in the next ring, and so on—you'll probably be surprised to find that the circle with the greatest damage for a hydrogen bomb is not 66.7 times bigger than the same circle at Hiroshima. In fact, the radius of severe-damage circle for a one megaton bomb is only 5.5 times bigger than the radius of the Hiroshima bomb's severe-damage circle, not 66.7 times. This is still a huge circle of destruction—something like nine miles across. But the point is that yield and destruction are not the same. Just because the yield increases by, say, one thousand times doesn't mean that that the area of destruction will increase by one thousand times. When you think differences in size with nuclear weapons, you have to keep in mind that yield is only loosely related to area of destruction.
[81] "Republicans lay battle lines over Biden's plan to retire B83 megaton bomb," *Defense News* (May 19, 2022), https://www.defensenews.com/congress/budget/2022/05/19/republicans-lay-battle-lines-over-bidens-plan-to-retire-b83-megaton-bomb/ (accessed May 24, 2023).
[82] The water slows the shell and reduces the cannon's range significantly. If the explosion occurs too close to you, the shock waves from it will do significant harm to your body.
[83] DeGroot, *The Bomb: A Life*, pp. 168–169.

6. THEY WEREN'T SHOCKED

It is possible because their greatest success is a myth.

He looked out the car window distractedly and sighed. It was Friday in the early part of August 1945, and Henry Stimson was tired.[84] The dapper and slightly imperious-looking Stimson was an incredibly hardworking and able public servant, who had served as President William Howard Taft's secretary of war from 1911 to 1913. He had been called back to be secretary of state for four years under President Herbert Hoover. And finally, he'd been in his current job as secretary of war (again) first under President Franklin D. Roosevelt and then, when Roosevelt had died earlier this spring, he'd continued on with Harry Truman, Roosevelt's successor. He didn't much like Truman, if the truth were to be told. Truman seemed a little like a country bumpkin to Stimson: not too smart, not at all sophisticated, and a little too willing to shoot from the hip. And Stimson was tired. He'd been working long hours for years now in jobs that carried a crushing weight of responsibility, and he needed a rest. Stimson was old. Seventy-eight was an advanced age for someone overseeing the vital and complex task of getting the United States to victory in World War II. He could feel the fatigue in his bones.

The United States and its allies had defeated the Germans in May, and that had lightened the load a bit. But here it was August, and they still weren't making much headway against Japan. Japan's leaders—despite the obvious hopelessness of their situation—refused to surrender. So, as the car sped on, Stimson thought of his beloved Highhold—a farm he owned on Long Island that he would sometimes sneak off to and spend a quiet weekend away from the cares of office. He'd made plans to fly up for the weekend and was on his way to the airport now.

But Stimson never got to his longed-for retreat. An aide who had raced to catch him dashed up at the airport with fabulous and startling news: Japan's leaders had signaled that they were willing—at last—to surrender.

* * *

The bombing of Hiroshima was the critical first impression of nuclear weapons. What the U.S. public, government officials, and military strategists learned from the episode was that nuclear weapons were "miracle" weapons that could force a stubborn opponent to surrender after only a few days. Their ability to destroy a city with a single blow made them extraordinary. And it was a kind of miracle: the United States bombed Hiroshima on Monday, it bombed Nagasaki on Thursday, and Japan's leaders signaled their willingness to surrender on Friday. What could be more convincing proof than that? And then it emerged that on August 15, when the emperor announced in a radio broadcast to the people of Japan that it was regretfully necessary to surrender, he said it was necessary because the enemy was now employing "a new and most cruel bomb." The U.S. public, government officials, and military strategists took the emperor at his word: His statement was proof that the Japanese had surrendered because of the atomic bombs.

For most Americans today, there is little doubt about what made Japan surrender. There are newspapers from that long-ago era whose headlines proclaim, "Japan Surrenders: Our Bomb Did It"—and most people still agree with those headlines. And because of that experience, they imagine nuclear weapons can win any war. That startling first impression of nuclear weapons has shaped thinking ever since.

What a first impression it was. These nuclear weapons had achieved what nothing else had been able to do: force Japan to surrender. And they'd done it in just four days. This feat was even more impressive because, up to that point, Japanese soldiers (and even civilians, for that matter) had displayed incredible courage and determination in battle—refusing to surrender despite insurmountable odds. The Japanese government, too, stubbornly refused to accept defeat, even though by all standard measures they were surely beaten: their fleet was confined to port, their air force was grounded for lack of fuel, most of their experienced pilots had been killed or captured, the majority of the islands they had taken in the Pacific had been recaptured, their homeland was surrounded by a submarine blockade that was keeping key war materials and food from getting to Japan, and their cities were being pounded by devastating air raids. All of this made the ability of nuclear weapons to force Japan to surrender seem even more amazing—not only could they force an adversary to surrender, but they could force even the most stubborn, determined, warlike adversary to surrender.

You can measure the strength of that first impression by listening to what people said. It wasn't long before they concluded that nuclear weapons had ushered in an enormous revolution in human affairs. A little less than two months after the bombings of Hiroshima and Nagasaki, President Truman said in an address before Congress:

> The discovery of the means of releasing atomic energy began a new era in the history of civilization. The scientific and industrial knowledge on which this discovery rests does not relate merely to another weapon. It may someday prove to be more revolutionary in the development of human society than the invention of the wheel, the use of metals, or the steam or internal combustion engine. Never in history has society been confronted with a power so full of potential danger and at the same time so full of promise for the future of man and for the peace of the world.... [I]n international relations as in domestic affairs, the release of atomic energy constitutes a force too revolutionary to consider in the framework of old ideas.[85]

The British Prime Minister, Clement Attlee, agreed, saying, "The modern conception of warfare to which in my lifetime we have become accustomed is now completely out of date."[86] When Winston Churchill (who preceded Attlee as British prime minister) first heard about the successful test of the United States' nuclear weapon, he said, "What was gunpowder? Trivial. What was electricity? Meaningless. This atomic bomb is the Second Coming in Wrath."[87]

The conclusion that nuclear weapons were revolutionary and enormously powerful was not just held by politicians. It was shared by experts as well. Bernard Brodie wrote that the "destruction ... from the Hiroshima and Nagasaki bombs ... left Japan and the whole world awed, appalled, and unequivocally convinced that a terrible new revolution had occurred in the means of waging war."[88] That long-ago consensus still holds today. Historian Gerard DeGroot, writing in 2011, noted that "[t]he atom bomb is rightly seen as a weapon that revolutionized war and diplomacy."[89] And international relations scholar Francis J. Gavin, writing in 2012, said, "The concept of the nuclear revolution—that the extraordinary power of these weapons has fundamentally and forever altered statecraft and calculations about war and peace—is widely accepted."[90]

The revolution changed not just the facts on the battlefield, but the way people thought about war. After all, these new weapons had "changed everything," according to Albert Einstein. Soon military officers, strategists, and academic experts were writing books arguing that new ways of thinking about war were required. They developed great chunks of new scholarship and new theories about how future wars would be fought. Perhaps the most famous of these was the "ladder" of escalation laid out by Herman Kahn, explaining why the use of one nuclear weapon would likely lead to the use of all of them.

These new ideas swept aside all objections. When there was a conflict between these new ways of thinking and old concepts, people tended to toss the old ideas overboard and stick with the new. For example, there had been city bombing before Hiroshima. But those attacks—using conventional bombs—had proved largely ineffective. No nation in World War II had surrendered because of city attacks carried out with conventional bombs. But from the point of view of these enthusiastic new strategists, that was no reason to doubt the power of nuclear weapons. Nuclear weapons changed the rules. Even if city bombing was ineffective in the past, with these new, revolutionary weapons, city bombing would surely be effective. They believed that old ideas had to be abandoned, old experiences forgotten. The nuclear era was a world made new, changed in almost every respect by the power of nuclear weapons. Even contradictions between these new beliefs and facts on the ground didn't overly concern them. After all, it would take time for our understanding to catch up with this new world—to see things in light of the new world that nuclear weapons created.

It was an intellectual revolution as well as a practical one. Historian Marc Trachtenberg summed it up this way: "The nuclear revolution was like a great earthquake, setting off a series of shock waves that gradually worked their way through the world political system."[91]

The sudden and surprising surrender by Japan at the end of World War II convinced almost everyone that nuclear weapons had extraordinary powers, unlike any weapon that had gone before. The events of 1945 created a reputation that nuclear weapons have had for more than seventy years and still have today. If belief in nuclear weapons were a religion, Hiroshima would be the first miracle.

HOW DID THEY DO IT?

There is still a bit of a mystery surrounding Japan's surrender. Obviously, nuclear weapons were remarkable. Obviously, they had won the war (people soon began to call them "the winning weapon").[92] Apparently, they

THEY WEREN'T SHOCKED

represented a revolution in war weaponry. But how did they do it? How did they win the war? They hadn't wiped out Japan's military forces; there were only a few thousand Japanese troops in either city. They hadn't killed all the leaders in one blow, leaving the survivors no choice but to surrender. After the bombings, there were still hundreds of thousands of Japanese troops, well supplied with ammunition, dug in near the beaches, waiting to repel the invaders. So, how did they do it? What gave the atomic bombs the remarkable ability to force an enemy to surrender in just days? And how could that power be harnessed and controlled in the future?

In the weeks and months after the end of World War II, government officials and military thinkers pondered what had happened. Then, in February 1947, Stimson—now retired—published what was, in essence, the official government explanation, the story of how nuclear weapons forced Japan to surrender. In an article in Harpers, he explained what had happened:

> We had developed a weapon of such a revolutionary character that its use against the enemy might well be expected to produce exactly the kind of shock on the Japanese ruling oligarchy which we desired ... the atomic bomb was more than a weapon of terrible destruction: it was a psychological weapon.[93]

And in Stimson's version of events, this psychological shock was what caused Japan's leaders to surrender.

> Hiroshima was bombed on August 6, and Nagasaki on August 9. These two cities were active working parts of the Japanese war effort. One was an army center; the other was naval and industrial. Hiroshima was the headquarters of the Japanese Army defending southern Japan and was a major military storage and assembly point. Nagasaki was a major seaport and it contained several large industrial plants of great wartime importance. We believed that our attacks had struck cities which must certainly be important to the Japanese military leaders, both Army and Navy, and we waited for a result. We waited one day.[94]

Stimson apparently believed, as did most people in the government, that these bombs were revolutionary. They were not only a remarkable advance in physics; they also provided a decisive advantage in war. This advantage was so great that, as Stimson said, "With its [the atomic bomb's] aid even a very powerful unsuspecting nation might be conquered within a very few days by a very much smaller one."[95] A few nuclear weapons, in other words, were worth whole armies of soldiers.[96]

And this first explanation—Stimson's framing of the nuclear weapons story—stuck. In the years since Stimson's 1947 article, the idea that nuclear weapons are psychological weapons—and extraordinarily powerful psychological weapons at that—has become one of the fundamental parts of the consensus about nuclear weapons.[97] But in the course of becoming the conventional wisdom, it was slightly altered. According to what became U.S. collective memory, it wasn't the bomb's ability to wipe out vital army and naval facilities that had won the war. People concluded that it was the shock of wiping out an entire city—including tens of thousands of civilians—that forced Japan to surrender.

The larger theme, however, that nuclear weapons are psychological weapons was soon firmly fixed in people's heads. And it is still believed today. It is a belief repeated by military men and defense department experts. It makes its appearance in scholarly articles and political debates. This notion of the unique shock value is at the heart of nuclear deterrence theory. We believe nuclear weapons can prevent attacks against us because we are convinced that they have an unusual psychological hold over our adversaries' minds. And the belief in the revolutionary ability of nuclear weapons to frighten adversaries is ratified by the number of nations that now rely on nuclear deterrence: the United States, Russia, the United Kingdom, France, China, Israel, India, Pakistan, and North Korea. And there are other states as well, countries that rely for their safety on nuclear weapons held by their allies, like Germany, Italy, Japan, South Korea, Belarus, and so on. All these countries have put their faith in the psychological power of nuclear weapons. The idea is central to the way the world is currently ordered.

THEY WEREN'T SHOCKED

Japan's leaders weren't so in awe of the bomb's remarkable power that they were coerced into surrendering. It should have been obvious to anyone who thought about the matter for even just a few minutes that they weren't shocked. The incredible psychological power of nuclear weapons wasn't "proved" by Japan's surrender. In fact, it's doubtful that any such psychological power played the slightest role in the surrender.

There were all sorts of clues that people in the United States missed in the rush and exultation of Japan's surrender. For example, the emperor's broadcast, which so many nuclear weapons advocates still point to today as "proof" of the bombings' impact, doesn't explain why Japan surrendered. The timing of events doesn't really make sense—it doesn't "fit." The reactions of Japan's leaders—the meetings they held and the plans they talked about—all seem to point away from Hiroshima as the cause of surrender, rather than toward it. And assigning the surrender to the atomic bombs ignores the elephant in the room: the fact that the war's character (and Japan's chances) changed radically the night before Nagasaki was bombed. That night, at midnight, the Russians, who had been neutral up to that point, suddenly renounced their neutrality pact with Japan, declared war, and launched a ferocious attack with more than 1.2 million men on parts of Manchuria, Sakhalin Island, and other territories.

BROADCAST

The emperor's radio broadcast is often pointed to as "proof" that it was the bomb that won the war. So, let's look at that a little closer. On August 15 throughout Japan, radios broadcast a prerecorded "rescript" (a special form of Imperial pronouncement) that was read by the emperor himself. Because of his semi-divine status, the emperor rarely appeared or spoke in public; so most Japanese citizens had never heard his voice before. It made a profound impact. The emperor explained that Japan had to surrender, saying, in part:

> Despite the best that has been done by everyone—the gallant fighting of military and naval forces, the diligence and assiduity of Our Servants of the State and the devoted service of Our one hundred million people, the war situation has developed not necessarily to Japan's advantage, while the general trends of the world have all turned against her interest. Moreover, the enemy has begun to employ a new and most cruel bomb, the power of which to do damage is indeed incalculable, taking the toll of many innocent lives.[98]

Nuclear weapons advocates repeatedly point to this statement and assert that it clearly explains why Japan surrendered. And for sixty years or so, that claim went unchallenged.

But it turns out that the emperor's radio broadcast is only half the story. The emperor also issued a second rescript, two days later on August 17, which was read out to every Japanese soldier and sailor, in every theater of the war.[99] This second message from the emperor also explained why Japan had to surrender. But this message—intended for military men—was quite different from the one broadcast on August 15:

> Now that the Soviet Union has entered the war, to continue under the present conditions at home and abroad would only

result in further useless damage and eventually endanger the very foundation of the empire's existence. Therefore, although the fighting spirit of the Imperial Navy and Army is still vigorous, I am going to make peace.[100]

So, while the Imperial rescript of August 15 referred obliquely to the atomic bomb and made no mention of the Soviet entry into the war, which would support the notion that the bomb was the reason Japan surrendered, the Imperial rescript of August 17 cited the Soviet entry, said it was decisive, and made no mention of the atomic bombs, which would make the Russian declaration of war the primary cause.

The most famous testimony, the testimony that supposedly proves that the atomic bombings forced Japan to surrender, is directly contradicted by the emperor himself. The two Imperial rescripts actually cancel each other out, proving nothing. They don't illuminate the emperor's state of mind or his reasons for advocating surrender. The only thing they show, perhaps, is that he was an astute leader who framed his arguments in terms his listeners would understand: When he talked to civilians (who mostly lived in cities), he talked about city bombing (the thing they would care about); when he talked to soldiers (who were naturally concerned with the fighting), he emphasized the military situation.

So, the piece of evidence most often quoted as proof that the bombings forced Japan to surrender, the emperor's rescript on August 15, essentially tells us nothing about why Japan's leaders decided to surrender.

TIMING

The second reason to doubt that Japan's leaders were shocked into surrendering is that the timing doesn't work. People who are shocked become emotional. They go aimlessly from place to place; they weep and tear their hair. And then, often, they do impulsive, emotional things. But when you step through the events of that fateful week carefully, it just doesn't hold together as a story of people who were shocked by the bombing of cities.

The United States bombed Hiroshima at 8:15 a.m. on Monday morning, August 6, 1945. Word of the bombing started to reach Tokyo within an hour. By the early morning of the next day (Tuesday), President Truman's announcement that an atomic bomb had been used was circulating in government circles in Tokyo.[101] On Wednesday, Foreign Minister Tōgō Shigenori met with the emperor, told him about the atomic attack, and urged that the war be ended. (Togo was the leader of the faction pressing for an immediate peace settlement.)[102] The emperor sent Togo to talk to

Premier Suzuki Kantarō, the head of the Supreme Council (then the effective ruling body of Japan), and Togo asked Suzuki to call a meeting of the Supreme Council. Suzuki, in turn, immediately sent messages to the four other members of the Supreme Council asking if they could meet in an emergency session to discuss the bombing of Hiroshima. Strangely, several said they were too busy. Then, at midnight on Wednesday night, the Soviet Union entered the war. At about 10:30 on Thursday morning, the Supreme Council met in an emergency session to discuss surrender. Nagasaki was bombed in the late morning soon after they sat down to talk. Later in the afternoon, the full Cabinet met; and finally on Friday, Japan signaled its willingness to surrender.

When western historians tell this story, the crucial date is always Monday, August 6. It is the dramatic high point—the climax—that the entire story builds toward.

But from the Japanese perspective, the important date was not Monday, August 6, but, rather, Thursday, August 9. After all, it was on this day that Japan's leaders, for the first time, sat down to discuss surrender. They had been at war for fourteen years, first with China alone and then with China plus the United States, Great Britain, and other allies. Fourteen years of war, but it was not until Thursday, August 9, that they seriously discussed surrender. From the Japanese perspective, Thursday is the pivot point. That was the day that everything changed.

So, what happened to motivate them to sit down finally and talk about giving in? What changed their minds? It can't have been Nagasaki. The members of the Supreme Council had already taken the extraordinary decision to meet in an emergency session to discuss surrender and were sitting at the table when news reached them that Nagasaki had been bombed. Nagasaki can't have been the reason they met.

It probably wasn't Hiroshima, either. That had happened three days previously, and they had already considered holding an emergency meeting to talk about Hiroshima the day before (at Tōgō's suggestion) and rejected the idea. So, what happened between Wednesday afternoon, when they were too busy with other things for an emergency meeting, and Thursday morning? What caused them to suddenly see the war in a whole new light?

What had changed between Wednesday night and Thursday morning was the Soviet Union's declaration of war at midnight. If you set aside the idea that Hiroshima must have forced Japan to surrender and consider the possibility that it was the Soviet declaration of war that did it, the timing and a whole host of other facts and events suddenly seem to fall into place. And from this new perspective, their decision to surrender looks more like a strategic calculation than an emotional reaction to a city being destroyed.

IT IS POSSIBLE

With the Soviet Union joining the war, Japan would now have to fight two great powers, not just one. The Soviets had the largest land army on the face of the earth at the time, and to make matters worse, the Soviets were going to invade from the north, while most of Japan's troops had been transferred to, and were dug in along, the southern coast, where the U.S. invasion was expected. In other words, the odds and the entire strategic situation had changed overnight. It had gone from difficult to hopeless, and in those circumstances, a decision to surrender was unavoidable—even for the stubborn and determined Japanese leadership.

Along with these strategic calculations, a powerful question of national identity also came into play. Japan's emperor and his semi-divine status had been a part of Japan's culture and politics for centuries. The Soviets, whose communist ideology prohibited religion, would almost certainly insist that Japan give up its national religion—at least in the portion of Japan's territory they occupied after the war. The Americans, on the other hand, with their liberal beliefs, might just be willing to let the emperor continue to play a role in Japan's national life. If Japan surrendered immediately to the United States, they might be able to preserve their emperor's status. Once the danger of Soviet conquest arose, the Japanese had powerful cultural and religious incentives to surrender quickly to the United States.

In light of this more careful review of the timing, it makes little psychological sense to argue that the reason for surrender was the bombing. The leaders of Japan had heard about the bombing, they had been asked to meet to talk about it (on Wednesday), they said they were too busy, and then the next morning they reversed themselves and decided to meet in an emergency session. Were they mentally unstable? Unable to make up their minds? Or did something happen on Wednesday night that changed the whole strategic calculus of the war?

EMOTIONAL TEMPERATURES

Is there evidence of the "shock" that the advocates of nuclear weapons talk so confidently about? If you look closely at the words of Japan's leaders in the days following the atomic bombing of Hiroshima, it is difficult to discern any such evidence. Kawabe Torashirō, deputy chief of staff of the army, wrote in his diary after the bombing that it was now clear that Hiroshima had been destroyed by an atomic bomb. This gave him, he said, "a serious jolt." But, he continued, "we must be tenacious and fight on."[103] Admiral Ugaki Matome wrote in his diary on August 7, "[I]t is clear this was a uranium bomb, and it deserves to be regarded as a real wonder making the outcome of the war more gloomy. We must think of some countermeasures against it immediately, and at the same time I wish we could create the same bomb."[104] Foreign Minister Tōgō heard about the atomic bombing of Hiroshima and went immediately to the Imperial Palace. He told the emperor that the bombing was an "opportunity" for pressing peace talks.[105] His response, in other words, was to wonder, "How can we use this bombing to reach a negotiated settlement of the war?" Two days after the bombing of Hiroshima, Admiral Yonai Mitsumasa, the navy minister, said to his friend, Admiral Takagi Sōkichi (and Takagi recorded the conversation in his diary), that although the domestic situation had "deteriorated" because of Hiroshima, what he was concerned about was another reduction in the rice ration, due to go into effect on August 11.[106]

Proponents of the Hiroshima narrative make much of the fact that the emperor sent numerous times for "more information" about the bombing once he had been made aware of it by Togo on Wednesday. But this isn't "shock"; it's asking for more information. Finally, consider what Minister of War Anami Korechika said. Anami was perhaps the most powerful man in Japan in August 1945—more powerful even than the emperor himself. Anami was not impressed with the destruction of Hiroshima and Nagasaki. On August 13, he said that the atomic bombings had been "no more menacing … than the firebombing Japan had endured for months."[107]

These reactions don't seem to be shocked reactions. There is determination ("we must be tenacious"), jealousy ("I wish we could create the same bomb"), opportunism (using the bomb in the effort to negotiate surrender), indifference (the belief that rice rationing was a greater threat to internal stability), a desire for more information (sending repeatedly for the latest reports), and utter disregard ("no more menacing … than the firebombing"). In fact, if you go through the diaries and meeting minutes, what is striking is what is not there. There are no descriptions of officials in tears at their desks as the realization sinks in that Japan has utterly lost the war.[108] There are no angry meetings with the participants shouting and trying desperately to think of a way to counteract this deadly new bomb. There are no handwringing pleas to the emperor not to surrender. All you have is a handful of lukewarm reactions—defiance, jealousy, indifference, and so on—none of which rises to the level of shock. Based on the evidence—on the diaries, meeting notes, and other documents—it is difficult to build a persuasive case that the Japanese high command was "shocked" by Hiroshima.

The fact that their words don't seem to betray shock is significant, but their actions are even more telling. If you compare what they did on the morning Hiroshima was bombed to what they did on the morning the Soviets declared war, the contrast is remarkable. On the morning after the Soviet Union came into the war, the leading officers of the Army rushed to headquarters for an impromptu emergency meeting to discuss how the war could continue now that the odds had shifted so decisively against them. Their desperation can be gauged by the suggestion of Deputy Chief of Staff of the Army Kawabe, who proposed that the Army kidnap the emperor and declare military rule.[109] And, of course, later that morning, Japan's leadership, in the form of the Supreme Council, met in an emergency session to discuss surrender for the first time in fourteen years.

Now contrast these three actions—leading military officers' emergency meeting, discussion of a coup, and an emergency meeting of the Supreme Council—with what happened on Monday, the day Hiroshima was bombed. Or rather, what didn't happen. There was no impromptu meeting of top Army officers on the morning Hiroshima was bombed. No one suggested overthrowing the government and declaring martial law on the morning Hiroshima was bombed. And, of course, Japan's government didn't meet in an emergency session to discuss surrender on the morning Hiroshima was bombed.

Viewed this way, it seems almost certain that what happened on Wednesday night was a crisis and what happened on Monday morning was not. If Japan's leaders were shocked by the bombing of Hiroshima, how could they have failed to act as if they were shocked? If Hiroshima was such a deep psychological blow, how did Japan's leaders maintain their normal routines, acting for the whole world as if nothing unusual was happening?

HISTORY

The strongest reason to believe that Japan's leaders weren't shocked is historical. The reason Stimson's explanation seems so unlikely is that it goes against the facts. It violates what amounts to a law of war that is ratified by thousands of years of precedent. In wartime, when the survival of the state is at stake, leaders are never shocked by civilian deaths—at least never shocked into surrendering. Never. There has never been an all-out war, one where the survival of the country was at stake, where the leaders turned to each other and said, "There are too many civilians dying. We have to surrender." Scour through the length and breadth of history, and you will find not a single solid instance of leaders surrendering when civilians died in large numbers.[110]

What you will find in that long historical record is a great deal of evidence—overwhelming evidence, in fact—that, when the survival of the nation is at stake, leaders don't take the deaths of civilians as strategically important. The bombing of London did not force Winston Churchill to go before the House of Commons and declare that Great Britain must now surrender. He was not coerced into calling for negotiations when Coventry was flattened. He may have wept when he toured areas that had been bombed, but he never considered ending the war because children and grandmothers were dying in the raids.

Adolf Hitler, early in the war, was concerned that Germans would lose heart if their cities were bombed. But even after Hamburg was mauled by firebombing, with an estimated thirty-seven thousand killed and more than 180,000 wounded, and great areas of the city were burned to cinders, Hitler never considered calling off the war. Even when the tide had turned decisively against Germany in 1945, and the Allies devastated Dresden, leaving most of the city center destroyed and killing an estimated twenty-five-thousand civilians, Hitler never wavered.

Joseph Stalin didn't consider surrendering when German forces surrounded and besieged Leningrad—a siege that was not lifted for 873 agonizing days. Starvation set in, and civilians began dying day after day. But Stalin carried resolutely on. Eventually, something like a million civilians died of hunger in Leningrad, but even this unimaginable death toll did not shake Stalin's resolve.

But the clearest example of the unimportance of civilians probably comes from the war in China. In the summer of 1938, Japan's invading armies had broken through Chinese defenses in the north and were advancing rapidly into central China. Many observers at the time believed that the Japanese might be close to winning the war. Chiang Kaishek, then China's leader, decided to blow the dams on the Yellow River so that the flooding would slow the Japanese soldiers' advance and give Chinese forces a chance to rally and set up new defenses. So, on June 9, 1938, the southern dike of the river at Huayuankou near Henan was breached and the water poured out.[111]

And the subsequent flooding did, in fact, slow the advance of Japanese forces. Arguably, this desperate act saved China from conquest. It also, however, drowned an estimated five hundred thousand Chinese civilians. Chiang, by his own order, killed half a million of his own civilians.

What this and many, many other examples make clear is that in a war where the survival of the state is on the line, civilian lives hardly matter. If Japan's leaders surrendered because they were shocked by civilian deaths, they were the first leaders in history to do so. But it hardly seems likely that they were.

JAPAN'S LEADERS

If any group of national leaders was able to ignore the horror of civilian deaths, it was the leadership of Japan during World War II. Perhaps only the leaders of the Mongols that swept out of the high Asian steppes and conquered the largest empire in history were more deeply steeped in a warrior's code of behavior.

As a practical matter, in World War II, Japan was being run by its military. Japan's constitution gave the power to run the country to a cabinet that had both military and civilian members in it. But by 1945, actual decision-making power had passed to a much smaller group of six. This Supreme Council consisted of the army minister, the navy minister, the army chief of staff, the navy chief of staff, the foreign minister, and the premier. So, of the six members of the council, only one was a civilian. (The Premier, Suzuki Kantarō, although he was serving in a civilian office, was a navy man through and through and had spent a lifetime serving Japan in uniform.) The military was firmly in control.

Culturally, Japan's leaders were especially determined and warlike. They were inculcated with the Japanese tradition of bushido—a warrior's code derived from the values of the samurai.[112] This same tradition was taught throughout the Army and Navy. To see how powerful it was, it is only necessary to examine how determined ordinary Japanese soldiers were and how rarely they surrendered. At Iwo Jima, for example, of the eighteen thousand Japanese soldiers who defended the island, only 216 were captured alive.[113] In other words, Japanese soldiers and sailors almost never surrendered.

It seems very unlikely that military men raised in such a stern tradition would be shocked by civilian deaths.

And remember, these same leaders had watched all summer long while U.S. bombers pounded Japanese cities without considering surrender. The U.S. Air Force began a series of massive air raids in early March 1945, destroying city after city—one every other day, on average. In all, sixty-eight Japanese cities were burned and destroyed that summer. It was the most ferocious campaign of city attacks ever launched in the history of war. But Japan's leaders did not see the civilian deaths and the destruction of so many cities and say to each other, "Now we have to surrender because too many civilians are dying."

Historian Herbert Bix offers a remarkable window into the thinking of Japan's leaders about city bombing. Describing events two days after the firebombing of Tokyo on March 10, 1945 (a raid that destroyed twelve square miles of the city and killed more than one hundred thousand people), he wrote:

> ... no less a person than retired foreign minister Shidehara Kijuro, once the very symbol of cooperation with Britain and the United States, gave expression to a feeling that was widely

held by Japan's ruling elites at this time: namely, Japan had to be patient and resist surrender no matter what. Shidehara had earlier advised Foreign Minister Shigemitsu that the people would gradually get used to being bombed daily. In time their unity and resolve would grow stronger, and this would allow the diplomats "room to devise plans for saving the country in this time of unprecedented crisis."

Now, on March 20, 1945, Shidehara wrote to his close friend Odaira Komatsuchi, the former vice president of the South Manchurian Railway Company, that, "[i]f we continue to fight back bravely, even if hundreds of thousands of noncombatants are killed, injured, or starved, even if millions of buildings are destroyed or burned," there would be room to produce a more advantageous international situation for Japan. With the country facing imminent absolute defeat, Shidehara still saw advantages in turning all of Japan into a battlefield, for then the enemy's lines of supply would become longer, making it more difficult for them to continue the war and giving diplomats room to maneuver. This was the mindset of … Shidehara; it was probably shared by [Emperor] Hirohito."[114]

And Shidehara was not one of the extremist ultranationalists in Japan's government. In wartime, civilians suffer. That is the attitude that leaders since time immemorial have held. Bix notes that "the Suzuki cabinet and the Supreme War Leadership Council never framed a peace maneuver from the viewpoint of saving the Japanese people from further destruction."[115]

To put the bombing of Hiroshima into perspective, if you graph the number of people immediately killed in all sixty-eight of the city attacks that spring and summer, Hiroshima ranks second. Tokyo, a conventional raid, killed more people. If you graph the square miles destroyed, Hiroshima is sixth. If you graph the percentage of the city destroyed, Hiroshima is fourteenth. If Japan's leaders had watched more than sixty cities bombed all summer long and not considered surrender, why would the destruction of one more city have suddenly shocked them into thinking that "too many civilians are dying"?

REALISM

So, if they weren't shocked, why did they surrender? They surrendered because they were realists. They knew the military situation by heart, and they knew that they couldn't possibly hold out now that Russia had come into the war.

IT IS POSSIBLE

Japan's military leaders knew that Japan was, at best, a middle-sized power and that, ultimately, they couldn't win a war against one of the great powers. They had measured the size and industrial output of the United States before they even started the war and knew that all-out victory wasn't possible. But they thought that even though they couldn't win, they might be able to hold on to some of what they'd gained. They had a plan for getting the United States to agree to let them keep some of the territory they'd conquered. Their plan was to dig in their best troops on the beaches where the Americans would have to invade, fight hard when the invasion came, and inflict so many casualties that eventually the United States would agree to a peace settlement that let Japan keep Korea, maybe, and part of northeast China, perhaps, and in the best case some of the islands in the Pacific. That was the plan.

But all of that was dependent on keeping the Soviet Union out of the war. Japan, with its tenacious military, might be able to hold off the United States if it fought on the defensive and concentrated its troops in the south of Japan, where they were sure the United States was going to invade. (The islands that the United States would use as the jumping-off point for the invasion, that it would use to base its planes, and that it would use to resupply the invasion force all lay to the south.) But if the Russians came into the war, they would be caught between two giant and remorseless jaws. The United States would hammer them from the south, and the Russians would come storming down out of the north.

For months now, they had been quietly stripping away the best units in China and on the northern islands of Japan and shipping them to Kyushu, the southernmost island of Japan, where they expected the Americans to attack. (And they had guessed right: the U.S. plan was to attack Kyushu.) Now their northern defenses were woefully thin. There was no way they could hold out if Soviet troops attacked in large numbers, and Japan's leaders knew it.

In fact, they had said as much in a meeting of the Supreme Council in June of that year. Thinking about the strategic situation, they concluded that the entire war effort depended on maintaining Russian neutrality. Soviet entry into the war "would determine the fate of the Empire," they concluded. General Kawabe, the deputy chief of staff of the army, expressed the consensus even more emphatically, saying, "the absolute maintenance of peace in our relations with the Soviet Union is imperative for the continuation of the war."[116]

So, there it is. Japan's leaders believed that they couldn't continue the war if the Russians decided to come into it. It was a strategic judgment made by experienced military leaders after months of careful consideration.

If the Soviet Union came into the war, they would have to surrender. And that is exactly what they did. Within thirty-six hours of the Soviet declaration of war, Japan signaled its willingness to surrender.

* * *

It is head-spinning to think that Japan didn't surrender because of the atomic bomb. The story that the atomic bomb won the war is so deeply woven into the fabric of who Americans believe they are that any other story seems ludicrous and wrong. But we cannot survive in a dangerous world stocked with nuclear weapons if we are unwilling to face up to reality. The reality—clearly and emphatically—is that Japan's leaders decided to surrender because the Russians declared war. They weren't shocked.

So, if Japan surrendered because of the Soviet Union's decision to join the war, where does that leave us? What difference does it make? Well, it hurts the pride (a little) of those of us who are Americans. We'd always taken complete credit for winning the war in the Pacific. Although to be fair, this doesn't change that much. Most of the work in the Pacific, most of the fighting, was done by U.S. troops.

The most important difference it makes is that it changes our assessment of nuclear weapons. It means that the crucial first impression about nuclear weapons was wrong. That first miraculous accomplishment—forcing a stubborn adversary to finally surrender when nothing else had worked—wasn't real. The miracle that inspired the founding of the nuclear religion was, it turns out, no miracle at all.

This means that the nuclear "revolution," the certainty that everything is now different, is also called into doubt. Viewed objectively, the nuclear revolution doesn't seem to have changed things that much. Countries still go to war, they still fight using tanks and rifles and artillery, strategy still works along familiar lines, and so on. For a revolution, there hasn't actually been that much change in the way things are done.

This raises a rather troubling question: how are we to think about all the years of belief in nuclear weapons? How can we recast our understanding of the world if these weapons failed to force Japan to surrender? One of the most important pillars that support the structure of ideas that make up nuclear weapons policy turns out not to be resting on bedrock but to be floating in midair. This means that the rationale for keeping nuclear weapons is much weaker than we ever imagined. It is possible that it is not persuasive at all.

[84] The three paragraphs that introduce this chapter are fictionalized history. They are based on my knowledge of the participants and the actual events. They are not factual history. I am not aware of an actual source that quotes Stimson describing Truman as unsophisticated. The interrupted trip to Highholds is real, however.

[85] S. David Broscious, "Longing for International Control, Banking on American Superiority; Harry S. Truman's Approach to Nuclear Weapons," in *Cold War Statesmen Confront the Bomb: Nuclear Diplomacy Since 1945*, ed. John Lewis Gaddis, Philip H. Gordon, Ernest R. May, and Jonathan Rosenberg (Oxford: Oxford University Press, 2005, ©1999), p.18.

[86] DeGroot, *The Bomb: A Life*, p. 224.

[87] Lamont, *Day of Trinity* (New York: Atheneum, 1985), p. 261.

[88] Bernard Brodie, *War & Politics* (New York: Macmillan, 1973), p. 54.

[89] Gerard J. DeGroot, "'Killing Is Easy': The Atomic Bomb and the Temptation of Terror," in *The Changing Character of War*, ed. Hew Strahan and Sibylle Scheipers (Oxford: Oxford University Press, 2011, p. 99.

[90] Gavin, *Nuclear Statecraft*, p. 158.

[91] Marc Trachtenberg, *History and Strategy* (Princeton, NJ: Princeton University Press, 1991), p. 146.

[92] Gregg Herken, *The Winning Weapon*, p. 7.

[93] Henry L. Stimson, "The Decision to Use the Atomic Bomb," *Harpers* (February 1947), p. 105.

[94] Henry L. Stimson, "The Decision to Use the Atomic Bomb," p. 105.

[95] Henry L. Stimson, "The Decision to Use the Atomic Bomb," p. 99.

[96] For a remarkable, concise, and (for me) mostly persuasive counterargument, see Admiral Sir Gerald Dickens, *Bombing and Strategy: The Fallacy of Total War* (London: Sampson Low, Marston & Co., 1947).

[97] Here, for example, is Gen. Kevin Chilton, former commander of STRATCOM, writing in *Strategic Studies Quarterly* in 2018: "Another fallacy is the notion that the deterrence mission can be adequately accomplished by substituting conventional warheads, because of their great accuracy, for nuclear warheads atop our ICBMs. Often referred to as a 'prompt conventional global strike' capability, the argument is that such weapons would be precise and in some cases powerful enough to destroy certain targets held at risk by today's nuclear forces. This argument does not appreciate the 'long, dark shadow' cast by the destructive power of nuclear weapons and the deterrent effect that 'shadow' enables. A nuclear warhead is terribly frightening; a 2,000-pound conventional warhead is not." Kevin Chilton, "Defending the Record of US Nuclear Deterrence," *Strategic Studies Quarterly* (Spring 2018), p. 13.

[98] Quoted in Frank, *Downfall*, p. 320.

[99] Robert J. C. Butow, *Japan's Decision to Surrender* (Stanford, CA: Stanford University Press, 1965), p. 225.

[100] Tsuyoshi Hasegawa, "The Atomic Bombs and the Soviet Invasion," in *The End of the Pacific War: Reappraisals*, ed. Tsuyoshi Hasegawa (Stanford, CA: Stanford University Press, 2007), p. 139.

101 Truman announced the bombing on Monday in Washington, but because of the time difference and delays in translating his statement, it wasn't until early Tuesday morning that word of the president's statement began circulating in Tokyo.

102 The leadership of Japan is often divided by historians into a "peace" faction and a "war" faction. This is misleading. All of Japan's leadership knew that the war would eventually have to end with some sort of negotiated settlement. They had known it since the outset of the war. Their goal was to conquer as much territory as they could during the war and then hope to retain some portion—but probably not all of it—when, inevitably, peace had to be negotiated. In this sense, both of the factions in the government were aiming to secure a peace treaty. One group wanted to do it now using diplomacy; the other wanted to fight off the U.S. invasion, inflict painful casualties, and then negotiate.

103 Tsuyoshi Hasegawa, *Racing the Enemy: Stalin, Truman, And the Surrender of Japan* (Cambridge, MA: Harvard University Press, 2005), p. 200.

104 Matome Ugaki, "Diary Entries of Admiral Matome Ugaki, August 7, 9, and 11, 1945," in *The Columbia Guide to Hiroshima and the Bomb*, ed. Michael Kort (New York: Columbia University Press, 2007), pp. 318–319.

105 Tōgō Shigenori, *The Cause of Japan* (New York: Simon and Schuster, 1956), p. 315.

106 Takagi Sōkichi, "Diary Entry for Wednesday August 8, 1945," Diary, https://nsarchive.gwu.edu/document/28493-document-67b-admiral-tagaki-diary-entry-wednesday-august-8-1945.

107 Forrest E. Morgan, *Compellence and the Strategic Culture of Imperial Japan: Implications for Coercive Diplomacy in the Twenty-First Century* (Westport, CT: Praeger, 2003), p. 216.

108 By contrast, when the emperor finally did convince them to surrender in the early morning hours of Friday the 10th, Japan's leaders—including the tough military ones—did weep.

109 This is less treasonous than it might sound to Western ears. Japanese history had a long tradition of powerful shoguns capturing the emperor in order to establish the legitimacy of their rule.

110 There is a legend of an Indian king in the time before Christ who stopped his campaigns of conquest because of the deaths of civilians, but it is no more than an unsubstantiated story found on a single stone marker from two thousand years ago.

111 Rana Mitter, *Forgotten Ally: China's World War II, 1937–1945* (New York: Houghton Mifflin Harcourt, 2013), pp. 157–164.

112 On the radical ideology that motivated some of Japan's leaders, see, for example, Walter A. Skya, *Japan's Holy War: The Ideology of Radical Shinto Ultranationalism* (Durham, NC: Duke University Press, 2009).

113 The National World War II Museum, The Battle for Iwo Jima, https://www.nationalww2museum.org/sites/default/files/2020-02/iwo-jima-fact-sheet.pdf (accessed May 25, 2023).

114 Herbert P. Bix, *Hirohito and the Making of Modern Japan* (New York: HarperCollins, 2000), p. 492.

[115] Bix, *Hirohito and the Making of Modern Japan*, p. 509.
[116] Quoted in Sadao Asada, "The Shock of the Atomic Bomb and Japan's Decision to Surrender—A Reconsideration," *Pacific Historical Review*, 67, no. 4 (November 1998), p. 504.

7. LUMPS OF COAL

It is possible because the reality is they're not very good weapons.

The fact that nuclear weapons have not been used since World War II is the most important fact about them—at least when it comes to understanding their nature. This remarkable non-use is a startling and unique circumstance for a weapon. What other weapon in history has experienced this strange absence of use? Can you think of another powerful weapon that people used once and then, for unexplained reasons, never used again? Clearly, the non-use of nuclear weapons is important. It means something. The question is, what?

It might mean that they are taboo, forbidden, and horrible in ways that prevent their use. It might mean that they have monstrous qualities that activate a kind of instinctive repugnance, that make us draw back in fear and disgust. Or perhaps it means that they are so godlike in their power, so awe-inspiring that we are afraid to touch them. In the presence of such massively destructive force, we are suddenly gripped by the feeling that we are too flawed, too inadequate to handle much less wield so much power.

But on the other hand, there is another thing it might mean. There is another possibility. Rather than being too big for us, it's possible we haven't used them because, in a sense, they are too small. Not actually small, but small in what they can do for us. It might mean that they are flawed; somehow, they are inadequate. It's possible we haven't used nuclear weapons because they are just not that useful. After all, that would fit the facts, too. You use a tool once, you see that it isn't adequate for the task at hand, and then you set it aside and use other tools instead.

Both theories—that nuclear weapons are too awesome to be used and that they aren't very useful—fit the facts. So, how can we choose between them? We could, of course, construct logic models or argue from first principles. But the realist approach, it seems to me, is to turn first to facts. So, to choose between these two possibilities, let's spend some time examining the historical record.

KOREA, DIEN BIEN PHU, VIETNAM, TAIWAN STRAITS

Nuclear-armed countries have been involved in wars a surprising number of times since World War II. The opportunity to use nuclear weapons has arisen again and again. And it's safe to say that in each of those wars, someone probably thought about using nuclear weapons. When the Soviet Union was bogged down in Afghanistan in the 1980s, for example, there must have been conversations in the Kremlin about whether the "ultimate" weapon could somehow get them out of that quagmire. But most of those discussions, if they did occur, remain secret.

There are, however, a few cases, episodes from the 1950s and a few more recent wars, where the questions were asked, the arguments were made, and the reasons why the weapons were not used have been recorded. They make fascinating reading. The image that exists in the popular culture of an all-powerful weapon that can do virtually anything is a far cry from the image that emerges from these debates among seasoned military officers, arguing strategy at the highest levels.

KOREAN WAR

The North Koreans invaded South Korea on June 25, 1950, and the campaign that followed favored first one side, then the other. On their own, the South Koreans were unable to hold the North's forces and were pushed back to the southernmost tip of Korea. When the United States and Allied forces (under the auspices of the United Nations) joined the war, the North's forces were repulsed, and U.S. and Allied forces rapidly retook South Korea and advanced through much of North Korea, getting almost to the North's border with China. At this point, China entered the war and turned the tables again, forcing U.S. and Allied forces to retreat roughly to the original dividing line between North and South Korea, the Thirty-eighth Parallel.

During these seesaw battles, the use of nuclear weapons was discussed inside the Truman administration more than once. When Truman mentioned in a press conference, however, that using nuclear weapons was under "active consideration," public opinion in the United Kingdom and elsewhere was very negative. Private criticism, too, was heard. The British Chiefs of Staff sent a discreet warning to General MacArthur, the U.S. commander in Korea:

> In our view, if the atom bomb were used in Korea, it would not only be ineffective in holding up the Chinese advance, but it would make the situation more desperate by inevitably bringing the Soviet Air Force into the war. The atom bomb is our ultimate weapon and we must keep it in reserve as a deterrent or for use in the event of the Russians launching a third world war.[117]

What is interesting about this note is the British Chiefs' belief that nuclear weapons would be "ineffective in holding up the Chinese advance." Here are military leaders at the highest level, just five years after Hiroshima, doubting the military utility of nuclear weapons. Of course, in 1950, the British had yet to test a nuclear weapon of their own, so perhaps one could argue that they lacked the requisite experience or didn't know what their U.S. colleagues knew and therefore lacked sound judgment about the military utility of nuclear weapons. But the fact is, the British Chiefs weren't the only experienced soldiers who reached that conclusion. Military men at the highest levels of the Eisenhower administration took the same view two years later.

Dwight D. Eisenhower, the general who had commanded all the Allied forces in Europe during World War II, had promised during the campaign that once he was president, he would end the war in Korea. After he was elected, in one of his first meetings with his national security advisors, Eisenhower said that he wanted to consider using tactical (i.e., smaller, battlefield) nuclear weapons in Korea to end the conflict there. He tasked them to explore the possibility:

> The outcome of these investigations was not particularly encouraging. Army Chief of Staff General J. Lawton Collins expressed himself as "very skeptical" about the military advantages; Chinese and North Korean forces were deeply entrenched along a 150 mile front, and recent bomb tests in Nevada had proven "that men can be very close to the explosion and not be hurt if they are well dug in." Paul Nitze, still serving as Director of the Policy Planning Staff, noted that if the weapons were not effective, their use might "depreciate the value of our stockpile." Certainly it would cause political difficulties with the allies. There was also the question of whether or not the Soviet Union might decide "to retaliate in

kind." If that happened, Collins added, American harbor facilities in Pusan and Inchon would make better targets for atomic bombs than would the enemy. Perhaps most disconcertingly, a National Security Council staff study concluded that it could not predict what the effect the use of atomic weapons would have on Communist forces in Korea, other than to demonstrate American determination.[118]

These various conclusions, all of them negative, are extraordinary. There seems to be some agreement that the bombs would be unlikely to do enough damage to Chinese troops to end the war, that they might incite the Soviet Union to join the war, and that if the weapons didn't end the war, it would undercut their reputation and reduce their value. The general public has been told again and again that nuclear weapons are the most powerful weapons in the world, yet the three experts quoted here—the army chief of staff, the director of policy planning at the State Department, and a national security council staffer—all believed that the military utility of nuclear weapons would be quite small. And remember that these men were not neophytes—they were the same experienced and knowledgeable men who had planned and carried out the vast U.S. military operations of World War II. They knew war. But they were each unable to be certain that nuclear weapons would be effective against North Korean and Chinese troops. The details of exactly why they believed this are still classified. But the conclusions are enough: top military and intelligence officials in the Eisenhower administration, after studying the matter closely, came to believe that using nuclear weapons against enemy troops in Korea not only might not be decisive but might not even provide a real "military advantage."

DIEN BIEN PHU

The second time the Eisenhower administration considered using nuclear weapons was during the collapse of the French colonial forces occupying Vietnam. Twenty thousand French soldiers were surrounded at a fortified position near a village in the northwest highlands of Vietnam called Dien Bien Phu. As the Viet Minh forces closed in, capturing outposts and setting up artillery emplacements that made resupply more difficult, the Eisenhower administration offered its French allies the use of nuclear weapons.[119] The plan called for U.S. forces to drop three tactical nuclear weapons on the encircling Viet Minh troops, thus breaking the siege.

In this case, we don't know what deliberations took place in Paris or what shape the discussions took. But we do know that whatever the risks were that went with the plan, they were so serious that the French apparently elected to face the possibility of sacrificing the entire twenty-thousand-man force rather than run those risks. In other words, they preferred catastrophe to trying to use nuclear weapons at Dien Bien Phu. And in the end, the stakes were very high. The Viet Minh overran the positions at Dien Bien Phu, and all the remaining French soldiers were either killed or captured. The defeat at Dien Bien Phu so shocked the French military, government, and people that France sued for peace directly after it fell.[120] What the French decided, apparently, was that they would rather lose the war than run the risks involved in having the Americans use nuclear weapons near their troops.

VIETNAM

A later study of the use of nuclear weapons, conducted during the time when the United States was fighting in Vietnam in the 1960s, similarly found little value in such an attack. JASON, an elite group of senior scientists that worked for the Department of the Air Force on emerging technology threats, conducted an in-depth study of using nuclear weapons in Vietnam. They concluded that the war there was not conducive to the use of nuclear weapons. "[T]he use of TNW [theater nuclear weapons, i.e., battlefield weapons] in Southeast Asia would offer the US no decisive military advantage if the use remained unilateral, and it would have strongly adverse military effects if the enemy were able to use TNW in reply."[121] Fighting a dispersed guerrilla-type force provided few inviting targets for nuclear weapons because enemy troops were rarely concentrated. The U.S. forces in Vietnam by that time, however, had set up supply depots, bustling port facilities, command centers, and other bases that would have made ideal targets for nuclear weapons if the Russians had decided to join the war or to supply their North Vietnamese allies with them.

Skeptics might respond to these examples from Korea and Vietnam by saying that they are from the 1950s and 1960s. They might say that since that time, nuclear weapons have been improved and the means of delivering them enhanced. They might argue that we simply don't know enough about the current capabilities of nuclear weapons to judge whether they would be effective. However, there is actually a surprising amount of evidence from a source with the kind of insider knowledge that puts an end to these objections.

GULF WAR

One of the strongest pieces of evidence comes from General (and then Secretary of State) Colin Powell. In the run-up to the Gulf War (the war to retake Kuwait after the Iraqis conquered it in August 1990), Secretary of

IT IS POSSIBLE

Defense Dick Cheney suggested that the use of nuclear weapons be explored. You can imagine what he was thinking: most of the fighting would be in the desert, and there were few civilians near the troops, so "collateral damage" (the unintended killing of civilians) would be kept to a minimum. If ever there were a place to demonstrate the military utility of nuclear weapons on the battlefield, this would be it. Powell, then chairman of the Joint Chiefs of Staff, tells what happened next in his autobiography:

> "Let's not even think about nukes," I said. "You know we're not going to let that genie loose."
>
> "Of course not," Cheney said. "But take a look to be thorough and just out of curiosity."
>
> I told Tom Kelly [a lieutenant general who was director of operations for the Joint Chiefs of Staff at the time] to gather a handful of people in the most secure cell in the building to work out nuclear strike options. The results unnerved me. To do serious damage to just one armored division dispersed in the desert would require a considerable number of small tactical nuclear weapons. I showed his analysis to Cheney and then had it destroyed. If I had had any doubts before about the practicality of nukes on the field of battle, this report clinched them.[122]

Powell says that this study, conducted in the run-up to the Gulf War, clinched his doubts about using nuclear weapons on the battlefield. What "a small number" was remains classified (as it should), but it was apparently enough to cause the existing copies of the report to be destroyed and the subject dropped.

The Eisenhower administration considered the use of nuclear weapons again and again—in Korea, at Dien Bein Phu, and (as we'll see in a minute) during the Taiwan Straits Crisis—but in each case, ultimately reached the same conclusion: there was little value in using them. Not because the weapons were too horrible. Not because it would have been immoral. But for pragmatic, practical reasons: The weapons would probably be ineffective militarily. As the historian John Lewis Gaddis explains:

> [T]here could be no assurance, whether in Korea, Indochina, or the Taiwan Strait, that the use of nuclear weapons would produce decisive military results. Their ineffectual use, moreover, might compromise the overall deterrent: if the bomb was seen to have no dramatic effect upon the North Koreans, the Chinese Communists, of the Viet Minh, then

how could it be expected to impress the Russians, or to reassure endangered allies? Better not to use it all, and thus preserve the credibility of a vague and, therefore, ominous threat.[123]

Although we don't know exactly what "unnerved" Colin Powell and the others who studied the possibility of using these weapons in the Gulf War, we do know that they ultimately reached a conclusion similar to the military officers who studied the question in the 1950s: There was no persuasive rationale for using battlefield nuclear weapons.

What we've learned from these episodes from U.S. history is that government officials decided (repeatedly) that it was better not to use nuclear weapons on the battlefield. And in several cases, their reason for not using them was a concern that using them would show how ineffective they could be. "But," someone might object, "these were, after all, wars of choice; they were wars waged by a great power against smaller powers where the survival of the great power was never at stake. What would happen," someone could ask, "if a nation that felt its survival was threatened had nuclear weapons at its disposal? If it was a choice of do or die, would nuclear weapons be used on the battlefield?"

ISRAEL

Again, this is a situation where we do not have to speculate. There is such an incident available for us to study. In 1973, Egypt and Syria attacked Israeli forces in the Sinai and in the Golan Heights—both areas that Israel had conquered in an earlier war. The attack took Israeli forces by surprise, and on the first day of the war, the Egyptians and especially the Syrians made real headway. Israel is a rather small country—you can drive from its top to its southern end in about five hours. So, the forces invading Israel—particularly the Syrian forces which came near to breaking through—clearly threatened the existence of Israel.

On the first day of the war, Defense Minister Moshe Dayan had visited the front and returned shaken and doubtful that Israel's forces could hold out against the Syrian onslaught in the Golan Heights. Reliable accounts say that early on the afternoon of October 7, as a meeting of the war cabinet in Prime Minister Golda Meir's office was drawing to a close, Dayan broached the subject of nuclear weapons. Warning that the situation was dire, he indicated that he had Israel's Director-general of the Atomic Energy Commission waiting outside the room and suggested that the prime minister and the rest of the war cabinet discuss options for exploding a nuclear weapon as a "demonstration" of the seriousness with which the

Israelis took the conflict. After some discussion, the prime minister decided that such options need not be explored at that time.[124] Since the tide of battle subsequently turned in Israel's favor, the matter did not come up again.

But these events raise a series of questions. Why did Dayan propose a demonstration instead of simply firing a nuclear weapon at Syrian forces? If nuclear weapons are such effective weapons, and if the survival of Israel was at stake, why not use the most powerful weapon at your disposal? Why did Prime Minister Meir eventually decide not to use nuclear weapons even as a demonstration? Why hold back a powerful weapon in a serious crisis? And finally, why did the Egyptians and Syrians think they could win a war against an opponent armed with nuclear weapons? Everyone was pretty sure that Israel had nuclear weapons—it had been reported in the *New York Times* in July 1970.[125]

What the story of the Middle East War of 1973 shows is that, in at least one case, a country that possessed the "most powerful weapon on earth" didn't use it—even when their survival as a country was in danger.

*   *   *

So, what's going on here? You've got these powerful weapons, but no one seems to want to use them. How could leaders and experienced military officers believe that such powerful weapons wouldn't be effective? What could possibly be wrong with such destructive weapons? How can we explain the distance between the popular view that nuclear weapons are "the ultimate weapon" and the professional military opinion that they wouldn't be useful on the battlefield?

EXPLOSIVES

The first reason that they aren't very useful is that they are weapons that explode. In the taxonomy of weapons, nuclear weapons fall into the larger class of explosives, and they inherit the limitations that explosives have. Explosions, compared to ground troops, are pretty limited in their application. They can't, for example, defend; they can't guard; they can't occupy; they can't reconnoiter; they can't arrest; they can't pacify; they can't capture territory; they can't make a show of force; and so on. Explosives do one thing: They blow up, destroying whatever is nearby.

Hitler could not have carried out the Holocaust with nuclear weapons. For that matter, he could not even have used large conventional explosives. The Jews of Europe were living side by side with non-Jews. Using large explosions to kill Jews would have also killed non-Jewish Germans (and Poles and Austrians and so on) in the hundreds of thousands or millions. This sort of undifferentiated slaughter would have done little to solidify Hitler's power in Germany. In fact, it would have undermined it. The Hutus could not have slaughtered the Tutsis in Rwanda using large explosions. Pol Pot could not have used explosions for the Killing Fields in Cambodia. And so on.

Explosives can't guard. Large explosives could not have surrounded Leningrad (now St. Petersburg) in World War II and kept the Russians in the city pinned down for years. Large explosions are bad weapons for capturing important transportation hubs. Defense often gathers and solidifies around important road junctions, like Gettysburg or Bastogne. If the Germans had possessed nuclear weapons (which they didn't) in World War II and had used a nuclear weapon on Bastogne, they would not only have achieved what they wanted—killing the soldiers controlling the road intersection—they would have created dangers for their own troops because all the troops marching through for days and weeks after would have been exposed to radiation poisoning.

Large explosions can't defend. Fast-moving vehicles racing across the border, soldiers infiltrating in small groups, convoys of covered trucks marked as medical supplies, paratroopers dropped behind your border in civilian areas—none of these can be effectively stopped with large explosions. And what is true for explosions is also true for nuclear weapons. Nuclear weapons can't guard, they can't capture, they can't defend, they can't do reconnaissance, they can't occupy, they can't capture enemy troops (or enemy leaders), and on and on. What they do is pulverize, make a hole, throw the dirt in the air, and leave lethal poison behind.

Nuclear weapons are the most destructive explosives there are. But that is not saying a great deal, because explosives are not a very useful category of weapon.

TOO BIG

Another drawback that explains why military leaders have hesitated to use nuclear weapons on the battlefield is that nuclear weapons are too big for most situations. Explosives are indiscriminate. If they create large explosions, those blasts destroy and kill everything within their radius of destruction—no matter what uniform the soldiers happen to be wearing. It's difficult, therefore, to use large explosives when their troops are near your troops. The battle I mentioned earlier, Dien Bien Phu, is a perfect illustration of this problem.

IT IS POSSIBLE

Dien Bien Phu is located in high hill country. The French forces had fortified a key road junction and some of the surrounding hilltops. The Viet Minh, using the heavy forest for cover, eventually encircled the entire French position and dug in. In some cases, the Viet Minh forces were less than a thousand yards from the French lines.

Most tactical nuclear weapons in the U.S. arsenal in 1954 had relatively large yields. The lowest yield of the MK-6 (a model that was produced in large numbers in the early 1950s), for example, was eight kilotons, or a little more than half the yield of the Hiroshima bomb. This would have given it a circle of maximum destruction about 0.7 mile across, with successively less lethal circles of destruction out to about two miles. The difficulty with using such a weapon at Dien Bien Phu was that the Viet Minh forces were so close to the French positions. If the United States had attacked the Viet Minh directly with nuclear weapons—aimed at their frontline soldiers, in other words—they would have also risked killing a significant number of French soldiers.

In addition, highly accurate missiles or even systems for global positioning had not yet been developed, so the bombers that dropped the bombs would have to be very accurate. Even the slightest error, or winds that slightly redirected the parachute that slowed the bomb's descent, could have resulted in thousands of French casualties or even the destruction of the entire position.

What makes Dien Bien Phu useful to study is that what happened there is not unusual in war. In fact, soldiers often fight at close quarters. Using a weapon that makes gigantic explosions when soldiers are close together almost ensures that you will accidentally kill some of your own soldiers. So, the simple size of nuclear weapons makes them unsuitable for a host of situations common in war.

DENSITY

Another reason nuclear weapons might be less effective on the battlefield than we imagine is an image in our heads. The popular image of enormously lethal weapons is skewed because of the way the damage from the weapon is typically presented. The power of nuclear weapons explosions is usually illustrated by a map with a series concentric circles of destruction drawn on it. The map is almost always of a city. You look at Chicago or New York or Los Angeles, see the circles of damage, and think "My God! All of lower Manhattan would be left in ruins! Think of those all those poor people." The mental image of death and destruction is intense and remains in your mind. And, no doubt, people present nuclear explosions in this way to impress the danger and destruction on you.

But the constant use of cities to show the effects of nuclear weapons seriously misrepresents their military value. The problem has to do with density. In a city, people are pretty tightly packed in; but a battlefield usually holds far fewer people than a city. Military units in the open field are an entirely different matter.

The Hiroshima bomb's area of maximum destruction extended out 0.8 miles from the epicenter. So, you could draw a circle on top of downtown Hiroshima 1.6 miles wide that represents the area where most buildings were destroyed and almost everyone died. Measured in blocks and skyscrapers, the effects seem enormous. But 1.6 miles doesn't represent the same amount of destruction as on a battlefield because cities are compact. People pile their dwellings one on top of the other and achieve extraordinary concentrations. In 2020, the average density of New York City was about 29,300 people per square mile.[126] Drop a Hiroshima-sized nuclear weapon on a city as dense as New York, and you can kill a lot of people.

Country dwellers, on the other hand, never achieve this kind of density. For instance, the average density in Nebraska in 2020 was 25.5 people per square mile, or roughly a thousand times less dense.[127] Randomly drop a Hiroshima-sized nuclear weapon in rural Nebraska, and the number of people who will probably be killed, on average, would be less than a school bus full.

Armies in the field are much more like the population of rural Nebraska than the population of New York City. Armies never achieve the kind of density that cities achieve. Military formations tend to spread out, distributing soldiers over the widest area possible in order to defend and hold the widest expanse of territory possible.

To take a very rough example, imagine that each division in an army was defending about fifteen miles of front.[128] For a division of fifteen thousand men, deployed in a typical battlefront of two regiments forward and one regiment somewhere in the rear, that works out to something like a density of 666 men per square mile.[129] If each nuclear weapon only killed 666 soldiers, that would be a lot of money and energy expended for very little gain. Of course, forces can be much more tightly packed, especially when they are attacking. But they could also be more thinly dispersed. And the results would vary widely depending on the height at which the nuclear weapon exploded, how deeply the soldiers were dug in, and many other variables. But this rough calculation gives a general sense of the scope of the problem.[130] Trying to use nuclear weapons on the battlefield is much more like nuking Nebraska than New York.

And before we move on to radiation, a quick word about terrain. Perhaps surprisingly, the terrain can have quite a large impact on nuclear explosions. And this may explain why the U.S. military leadership was so skeptical about using nuclear weapons in Korea. Consider the terrain of Korea, which consists, in many places, of steep hillsides and deep ravines. If a brigade is occupying a valley and you use a nuclear weapon to attack it, you can probably do some damage to the troops there. But the brigade in the next valley over and the one in the valley on the other side will be largely unaffected by the blast. The steep slopes will reflect the force of the blast upward. The results of bombing the Japanese city of Nagasaki in World War II demonstrate this. Nagasaki—a city divided between two valleys with a high ridge between—suffered much less damage and destruction than Hiroshima. Because of heavy cloud cover, the bomb intended for Nagasaki was dropped considerably off target. Although the part of the city directly under the atomic blast was devastated, the eastern part of the city was left largely intact, sheltered by the high ridge that lay between.

RADIATION

The dispute over the islands of Quemoy and Matsu that took place in 1954, often referred to as the first Taiwan Straits Crisis, illustrates another problem with nuclear weapons. When the Nationalist Chinese, led by Chiang Kaishek, retreated to the island of Taiwan at the end of the Chinese civil war, leaving the Communists in possession of mainland China, they left troops on a series of small islands near the Chinese coast. These tiny outposts were close enough to the mainland that they were frequently shelled, and periodically there were concerns that the Communist Chinese were going to launch an invasion to retake them. When tensions flared in 1954, the Eisenhower administration explored ways to either defuse the situation or, on the other hand, be sure to win if a military conflict resulted. One option they explored was the use of nuclear weapons against targets in mainland China. But this raised difficulties.

For one thing, if nuclear weapons were used against targets on the mainland, the disproportion of the attack would likely lead to a negative world reaction. Defending relatively small islands by launching attacks with huge explosive devices might seem disproportionate to some people: why respond to such a tiny provocation with such a massive response? But there was another problem.

> Secretary of State [John Foster] Dulles saw still other difficulties. Might not Chiang Kaishek himself have the feeling "that an atomic attack on the mainland as a beginning would

be a poor way to gain the support of the Chinese people for his cause?[131]

How, Dulles asked his subordinates, could the Nationalist Chinese ever reconquer China if their ally—the United States—was slaughtering civilians there with nuclear weapons? And, there was another, potentially more serious problem.

> Central Intelligence Agency Director Allen Dulles warned the White House early in April that it might be difficult to use such devices against military emplacements on the mainland without subjecting Quemoy itself [one of the islands] to dangerous fallout; furthermore, "if the winds were wrong, the fallout would endanger the city of Amoy [on the mainland—called Xiamen today] with its several hundred thousand civilian residents."[132]

Eisenhower eventually framed his conclusion this way: "I do not think that it would be wise, unless we were forced to do it, to atomize the Mainland opposite them [the Nationalist-held islands]."[133] The president was reluctant to launch an attack that would release so much harmful radiation.

In the list of inherent limitations of nuclear weapons, perhaps the most important is radiation. When a nuclear weapon explodes, it leaves a poison—radiation—at the site of the explosion and that can also be carried far downwind. Radiation is deadly near the site of the explosion, with less fatal effects the farther you move away from the point of the explosion. Small radioactive particles are created by any nuclear explosion. Some of these particles settle on the surface of fields, buildings, or any other exposed surfaces near the explosion. But others attach themselves to dust and dirt thrown up by the explosion, dust that can be borne by the wind for considerable distances—some can even travel to the other side of the globe. Low doses of radiation can be harmless, but they can also eventually cause cancer, and larger doses can lead to immediate radiation sickness or death. From a military point of view, the radiation spewed out by nuclear weapons is more than an inconvenience; it is an undeniable danger.

Using a weapon that releases poison into the winds complicates military planning. For one thing, wind introduces an uncomfortable uncertainty into military calculations. A wind that is favorable when the attack is planned or approved (i.e., one that blows toward the enemy's troops and homeland) may veer around suddenly just before, during, or after an attack and spread dangerous poison onto your own position, troops, and supplies. This uncertainty raises enormous difficulties. It will rarely be possible in a fast-moving battle to wait for a favorable wind before attacking.

To get a sense of the size of the problem, look at this map of an attack scenario in Germany during the Cold War where 171 nuclear weapons are used. The map shows the areas that would be subjected to harmful levels of radiation poisoning, assuming typical wind patterns. Even this relatively small number of weapons would spread potentially lethal doses of radiation over much of the country.

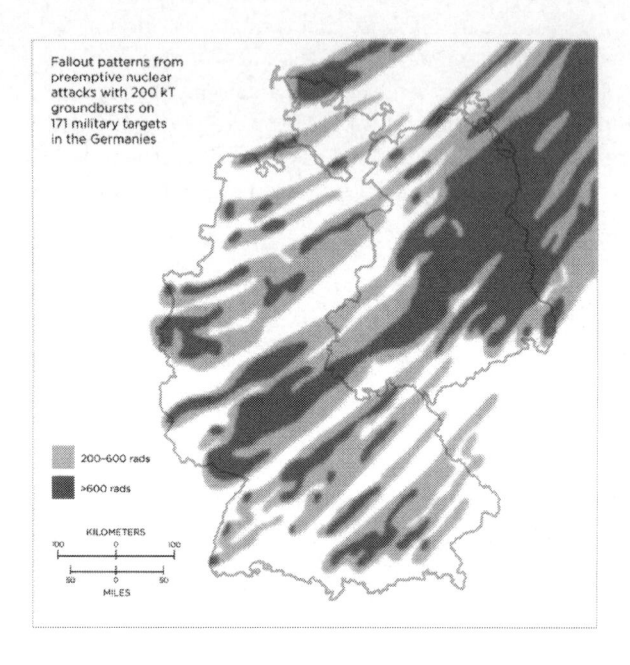

Fallout patterns from preemptive nuclear attacks with 200 kT groundbursts on 171 military targets in the Germanies

200–600 rads
>600 rads

Imagine trying to handle the disruption that such an attack would cause. Any troops that move through the area where nuclear weapons were used are likely to be affected by radiation. Dislocations of civilian populations, transportation, basic utilities, and so on would make operating in this theater extremely challenging. Even troops that remain in place after the attack could be affected if the wind suddenly changed direction and sent these radiation plumes back over them.

So, one of the principal challenges of using nuclear weapons in war is trying to figure out how it might be possible to work around the difficulties created by radiation. Of course, troops can be outfitted with special protective suits, and these can be pretty effective against radiation. But the troops' fighting effectiveness will be significantly degraded. Their movements and perceptions will be inhibited. And care will have to be taken with any object that may have been exposed to the dust, making the handling of ammunition, food, tents, medical supplies, and all the other accouterments of modern warfare difficult and dangerous. Even a foxhole that has been dusted with radioactive particles can be a danger to its occupants. Since radiation can only be detected by special equipment—it cannot be smelled, tasted, seen, heard, or felt—working in an environment where radiation may be present creates untold difficulties.

A nuclear battlefield requires that you either institute cumbersome protection and decontamination procedures—and therefore let military efforts grind to a standstill while you wipe and clean—or be willing to lose some of your own soldiers to radiation poisoning. In effect, using nuclear weapons on the battlefield means killing some of your own soldiers.

During the 1950s, scientists apparently believed that they could solve the problem of radioactive fallout. Secretary of State John Foster Dulles, in a press release about nuclear weapons, explained that the testing that was then underway was "not designed … to demonstrate the capacity to build bigger bombs" but rather to develop cleaner bombs. The United States had plenty of big nuclear weapons. "The whole problem," he said, "is to develop the capacity to have smaller weapons with more tactical power and with less fallout, so they become more distinctly a military weapon than a mass-destruction weapon."[134] It appears Dulles was confident this could be done.

But over time, it has become clear that the scientists who were advising Dulles were wrong about their estimate that they could one day build nuclear weapons with less radiation. Despite what they may have said to Dulles about being on the edge of building "clean bombs," no such weapons have ever been developed. Nuclear explosions create radiation every time. It is one of the enduring characteristics of the weapons. And that makes them very difficult to use on any battlefield in a useful way.

The radiation that nuclear weapons create—that they always create—will likely kill at least some of your own soldiers or civilians.

MILITARY OPINION

If you had to guess, based on what active military officers say in public, you'd probably say there's nearly unanimous support among military officers for nuclear weapons. But public statements by serving officers can be deceptive. Most military personnel feel obligated to support the policy of the United States—whatever it is—while serving. So, as long as it is the policy of the U.S. government to rely on nuclear weapons, most people in the military will not express objections or doubts about that choice publicly.

But the history we've reviewed here points to considerable doubts about the utility of nuclear weapons. And a second source of information points to even greater doubts. A surprisingly large number of military officers have publicly denounced nuclear weapons once the officers retired. Literally hundreds of high-ranking military officers have, at one time or another, expressed strong reservations about the weapons and sometimes called for their elimination.[135] These include officers with intimate experience with nuclear weapons and the highest levels of access to classified information. Perhaps the most prominent example of someone who criticized nuclear weapons and knew what he was talking about was General Lee Butler, the four-star Air Force general who was in charge of Strategic Air Command (SAC) in the 1990s and then became the first commander of Strategic Command (STRATCOM) when the three separate nuclear commands—missile, bomber, and submarine—were combined. After his service to the United States was finished, Butler said, in remarks at the National Press Club in Washington, D.C.:

> Is it possible to forge a global consensus on the proposition that nuclear weapons have no defensible role; that the broader consequences of their employment transcend any asserted military utility; and that as true weapons of mass destruction, the case for their elimination is a thousand-fold stronger and more urgent [than] that for deadly chemicals and viruses already widely declared immoral, illegitimate, subject to destruction and prohibited from any future production?
>
> I am persuaded that such a consensus is not only possible, it is imperative.[136]

When you consider how many times nuclear weapons were not used—even during wars where American servicemen and women were dying—it has to raise real doubts about their utility. Add to that the many strong statements by retired military officers and it starts to seem as if there is a considerable proportion of military personnel who do not believe in the military utility of nuclear weapons.

EUROPE

Perhaps the most damning indictment of the military utility of nuclear weapons comes from a man who never publicly criticized them. Born into wealth and privilege, mild-mannered George Herbert Walker Bush was an insider all his life. Two words that could describe him well are "conventional" and "successful." When World War II came, he signed up to join the Navy on the first day he could: his eighteenth birthday. He flew a Grumman TBF Avenger torpedo bomber, saw action in the Pacific, and was awarded the Distinguished Flying Cross.

After the war, he worked in finance and founded a successful oil company. In 1966, he was elected to Congress as a Republican. During the Nixon and Ford administrations, he served in a number of high-level positions: as ambassador to the United Nations, as chairman of the Republican National Committee, and as director of the Central Intelligence Agency. He seemed a natural choice to be Ronald Reagan's vice present in 1980 and served two terms before being elected president in his own right in 1988.

Bush was a moderate Republican, an establishment man, who rarely surprised people, rarely took risks, and rarely made unexpected choices. Comedians who mimicked him captured the essence of his public persona by having him say the phrase "wouldn't be prudent" again and again, which makes the decision he made on nuclear weapons all the more remarkable and, frankly, astonishing.

In the 1950s, when the democratic governments of Europe were still recovering from World War II, the military forces of those governments were necessarily weak. The Eisenhower administration decided to transfer tactical nuclear weapons to Europe rather than send hundreds of thousands of American soldiers to stand guard there for decades to come. Over the years, more and more nuclear weapons were sent to Europe—eventually more than seven thousand. The doctrine of using nuclear weapons to defend Europe was enshrined in U.S. policy and became a part of NATO policy, too. There was widespread support for keeping nuclear weapons in Europe in the military, among diplomats, and in government.

But despite this widespread support, some argued that the weapons made little actual sense. The physicist Freeman Dyson, for example, who held a high-level security clearance to advise the military on possible future military technologies, wrote scathingly about plans to defend Europe with nuclear weapons:

> [I]t remains true that the doctrines governing the use and deployment of tactical nuclear weapons are basically out of touch with reality. The doctrines are based on the idea that a tactical nuclear operation can be commanded and controlled like an ordinary non-nuclear campaign. This idea may have made sense in the 1950's, but it certainly makes no sense in the 1980's. I have seen the result of computer simulations of tactical nuclear wars under modern conditions, with thousands of warheads deployed on both sides. The computer wars uniformly

end in chaos. High-yield weapons are used on a massive scale because nobody knows accurately where the moving targets are. Civilian casualties, if the war is fought in a populated area, are unimaginable. If even the computers are not able to fight a tactical nuclear war without destroying Europe, what hope is there that real soldiers in the fog and flames of a real battlefield could do better?[137]

Dyson's objections, and similar criticisms, were not part of the policy mainstream and hardly seemed likely ever to become policy—that is, until George H.W. Bush did a remarkable thing.

In 1991, the president announced that the United States was retiring almost all of its tactical nuclear weapons, not only from Europe but from around the globe. He didn't do this as part of a treaty. He didn't do it because there was a quiet deal with the Soviets that they would do the same. He did it unilaterally. Imagine a careful, cautious man, given to prudence and the middle of the road, drastically reducing the nuclear weapons the United States had in the field unilaterally.

When an establishment man turns his back on the weapons that are supposedly the central element in American national security, it is surprising, confusing, and shocking. What reason could Bush have had for retiring these weapons that had been endorsed by generations of security experts? Bush wasn't an anti-nuclear crusader. He wasn't some liberal hoping to make a name for himself. Bush retired the nukes quietly, without fanfare. He wasn't looking for a political payoff, hoping to suddenly win the votes of the political left. He didn't do it as part of a diplomatic deal that reaped benefits for the United States. Since all the other possible motivations don't add up, we're forced to conclude that Bush retired the tactical weapons for the simplest reason of all: they weren't very good weapons.

And that impression, that intuition that Bush felt secure in retiring the weapons because they are difficult to actually use in war is confirmed by the fact that the Soviets, shortly afterward, followed suit. They retired a good portion of their tactical nuclear weapons as well. They apparently agreed with Bush that there was something so seriously wrong with battlefield nuclear weapons that most of them could safely be scrapped.

The unlikely story of the way so many U.S. nuclear weapons were taken out of service is rarely told. But it is clearly the final proof that leaders and military officers have profound doubts about trying to use nuclear weapons in battle.

EXPLOSIONS USED AGAINST COUNTRIES

"All right, so maybe there are drawbacks to using nuclear weapons on the battlefield, but surely dropping bombs on your adversary's cities and wiping out his population is a way to win the war, right?" someone might ask.

The problem with homeland attacks is that homelands don't make war; soldiers make war. Of course, homelands supply the weapons and food that soldiers need to fight; but the notion that attacking homelands is an effective way to win a war misunderstands what war is. War is, first and foremost, a contest between soldiers, and if you fail to keep this in mind and get distracted by landing punches on civilians far from the front lines, you are likely to lose.

Consider the scenarios that worried U.S. military planners in the years after World War II. The Soviet Union had millions of men under arms, poised on Europe's borders. If the United States had launched a nuclear attack against Soviet cities, even if those attacks had killed every civilian in the entire country, the Soviet armies could still have gone crashing forward into Europe, sweeping aside the relatively light defenses, capturing supplies as they went, and no doubt occupying the entire continent before U.S. land forces could be shipped across the Atlantic to oppose them.

Ulysses S. Grant is famous for winning the battle of Vicksburg by crossing the Mississippi, cutting himself off from supplies and reinforcements, living off the land, and eventually defeating the Confederate forces with the supplies he could carry with him. He voluntarily cut himself off but won anyway.

War is about defeating the enemy's military, not punishing people who are unable to fight back. Sometimes destroying manufacturing facilities for vital military supplies can help win wars—particularly in wars of attrition. But by and large, it is better to take aim at the soldiers in the field than economic and civilian targets far behind the lines. Even used far from the battlefield, nuclear weapons have drawbacks and problems that are often overlooked.

ATTACKING ISRAEL

One way to understand the difficulties of attacking homelands is to consider an imaginary scenario in which Iran attacks Israel with nuclear weapons. Of course, Iran does not currently have nuclear weapons, and even if they had them, the likelihood of their launching a nuclear attack on Israel is remote. But people often talk about such an attack, so let's use it as our example. As it happens, it throws significant light on some of the weapon's drawbacks.

IT IS POSSIBLE

Imagine that Iran had developed a small arsenal of about thirty nuclear weapons. Imagine that it decided to use all thirty simultaneously in an attack on Israel. Of course, such an attack would probably call forth a severe response by Israel's allies, and possibly even by Israel itself. (Israel reportedly has an arsenal of nuclear weapons.) But it is a measure of how serious the drawbacks of nuclear weapons are that, even before considering the possibility of retaliation, the weapons themselves create almost insurmountable obstacles.

The first problem is that there is a large Palestinian population that is close to Israel's population. Because Israel is so small and the footprint of each nuclear weapon is so large, almost inevitably any nuclear attack against Israel would kill tens of thousands of Palestinians along with Israelis. Even though there is sometimes said to be no love lost between Persians and Arabs, Iran still has ambitions to become an important state in the Middle East and to become the leading proponent of Islam in the world. Neither of these ambitions would be helped by slaughtering thousands or even tens of thousands of Palestinians.

The second problem with attacking Israel with nuclear weapons is Jerusalem. No strike against Israel could be considered thorough unless Jerusalem were attacked. It is the symbolic heart of the Jewish state, site of the remains of the Temple, one of the most important sites in the Jewish religion. However, there is a problem. The third holiest shrine in Islam is also located in the heart of Jerusalem. Using nuclear weapons against the city would mean destroying the Al-Aqsa Mosque on the Temple Mount. Islam is a religion in which holy shrines—physical locations—are particularly sacred. Any nation that destroyed the third holiest shrine in Islam would be reviled by Muslims worldwide. If Iran destroyed the third holiest Islamic shrine, it would deeply damage or destroy any legitimacy their claim to Islamic leadership might have.

Finally, the third problem is that the Middle East is a small place and the wind blows. Depending on a number of factors, including the targets of the blasts, the height of the explosions, and the direction of the wind at the time, an attack on Israel could easily result in radiation plumes that left thirty thousand dead in Amman, Jordan, or 150,000 dead in Cairo, Egypt. Depending on the strength of the wind, the fallout could even reach as far as Iran itself. Again, killing tens or hundreds of thousands of civilians in neighboring Arab countries would deeply damage any chance Iran had of increasing its political influence in the Middle East. And all of these problems result just from the use of the bombs. These drawbacks don't even take into account the possibility of worldwide condemnation, international economic isolation, or conventional or nuclear counterattack by Israel or its allies.

It is useful to compare and contrast such an attack with a conventional attack on vital Israeli military forces and command centers. An attack with conventional missiles could, in theory, do significant damage to Israeli military capacity and would provide political benefits for Iran—increasing its standing with many states in the Middle East that see Israel as an enemy. Such a surgical, limited strike would carry none of the costs listed above—killing tens of thousands of Palestinians, destroying holy shrines, or killing civilians in other countries with radiation. Of course, Iran would still face the possibility of retribution by Israel and its allies, but on the whole, in these circumstances, any sensible military commander would recommend against a nuclear attack if a precision attack with conventional weapons were available. And even if a conventional attack were not available, a sensible military commander could still make a strong case in favor of not attacking at all rather than using nuclear weapons. In a side-by-side comparison, nuclear weapons have remarkable drawbacks.

These drawbacks are not dependent on the local military balance or international alliances. In a world in which Iran had a monopoly on nuclear weapons, all these same drawbacks would still hold. This leads to a rather dramatic conclusion about the flaws inherent in nuclear weapons: even if a state has a monopoly on nuclear weapons, the weapons carry such serious impediments to sensible military employment that in almost all situations, they are virtually useless.

A SURGICAL STRIKE

A second example of the unexpected difficulties that can arise from homeland attacks comes from a famous study by two physicists in response to something the secretary of defense said at the time.

In the early 1970s, then Secretary of Defense James Schlesinger argued that a limited Soviet strike against U.S. nuclear forces was a plausible scenario. It presented, he said, a realistic danger that the United States had to take into account in its thinking about nuclear weapons. (Unsurprisingly, the solution to the problem he was proposing involved spending more money and building more nuclear weapons with more capabilities.) The Soviets might believe, Schlesinger argued, that they could launch a strike limited to U.S. nuclear forces, and because U.S. civilian casualties would be relatively low, the United States would not strike back. He argued, in other words, that U.S. forces were vulnerable to a nuclear "surgical strike."

In Congressional testimony on March 4, 1974, Schlesinger said that the Russians could expect that the United States would not respond with a nuclear retaliation because the number of civilians killed would be relatively small—in the neighborhood of "hundreds of thousands" rather than "tens or hundreds of millions."[138] On the surface, this number might not seem unreasonable. The United States carefully placed its missile silos far from any population centers in prairie states. If you didn't think closely about it, you might be able to imagine nuclear warheads striking these silos far out on the plains and most of the casualties coming from the attacks on submarine bases in harbors near populated cities.

But the claim didn't sound right to Frank von Hippel and Sidney Drell, two respected senior physicists with considerable expertise in thinking about nuclear weapons. (Von Hippel subsequently served in a White House staff position working on national security issues.) They both knew nuclear weapons well and the notion that a large number of these gigantic, clumsy weapons could be detonated and only cause a relatively small number of casualties seemed intuitively wrong to them. They set to work, digging into the scenarios, and subsequently published their findings in *Scientific American*.

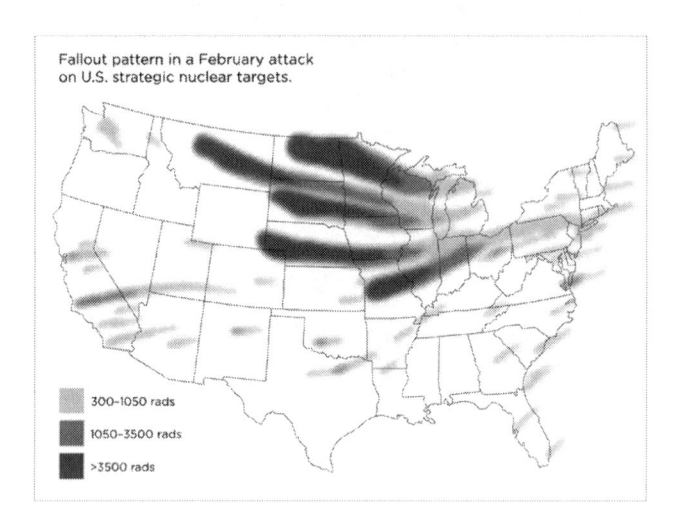

Fallout pattern in a February attack on U.S. strategic nuclear targets.

300-1050 rads
1050-3500 rads
>3500 rads

The results showed that even the most carefully designed "surgical" attack—strictly limited to nuclear silos, air bases, and submarine bases—could potentially kill eighteen million people.[139] The problem was not just the enormous size of the explosions but also their radioactive fallout. Even though most of the bombs would fall on isolated missile silos, the radiation from those explosions, carried downwind, would be deadly to millions of people. In light of the Drell/von Hippel study, the argument that Schlesinger was trying to make evaporated. It was clear that even a limited strike on U.S. nuclear forces could lead to massive casualties that would almost certainly call forth a full-scale U.S. response.

A DIFFERENT WORLD

Finally, the strongest reason that nuclear weapons can't win wars is an argument that things have changed. When the world consisted of two superpowers—the United States and the Soviet Union—it might have been possible to convince yourself that a nuclear war could be "won." (Although President Kennedy once told a columnist friend that "[o]nly 'fools' could cling to the idea of victory in a nuclear war.")[140] But in that bygone world, you could tell yourself that even though both sides would be incredibly devastated, one of them might still rise to victory.

The 1950s and 1960s were a time when the United States and the Soviet Union stood like giants in a world of Lilliputians. The two countries towered high over all others. The countries that had been fought over in World War II were devastated, prostrate, and incredibly weak. Others had yet to develop. In such a world, it might be possible to imagine that both giants could suffer from a war. But they would still be more powerful than any of the Lilliputians, and whichever one of them recovered faster might, in some sense, have "won" the war. I have grave doubts that such a "victory" is possible. But it is a fact that it was seriously discussed in strategy and government circles in the 1960s.

But today, such an outcome is simply not possible. All the Lilliputians have grown, and new giants have arisen. If the United States fought a war with Russia, for example, the world after the war would be dominated by China. China is currently so powerful that if the United States and Russia were devastated, it would stand like a colossus over the rest.

And if the United States fought a nuclear war with China, Russia would probably move aggressively to conquer first Europe and then significant portions of the rest of the world. And if Russia, China, and the United States all fought a three-way nuclear war, then Europe or Brazil or Nigeria would end up as the dominant power in the world. There are too many nations with the population and strength to imagine themselves leading a post-nuclear war world for any such conflict to end in anything but a disaster for the participants.

As General Douglas MacArthur argued in 1961: "Global war has become a Frankenstein to destroy both sides. No longer is it a weapon of adventure—the shortcut to international power. If you lose, you are annihilated. If you win, you stand only to lose. No longer does it possess even the chance of a winner of a duel. It contains now only the germs of double suicide."[141]

IT IS POSSIBLE

In a bipolar world, dominated by two giants, people might have been able to convince themselves that a war between those two might eventually be won by one or the other. In a world with many powerful nations, being devastated by nuclear war is the end of whatever world leadership your country once enjoyed. In fact, you'll probably end up dominated, if not actually conquered, by some other country. You will end your days impoverished, hungry, and no longer the masters of your own fate. Fighting a nuclear war is a prescription for slavery, starvation, or both.

This means that "nuclear war" is an imaginary concept. War, by definition, is a contest that can be won. Nuclear war cannot be won in any reasonable sense. Therefore, nuclear war is not war. It is slaughter or self-immolation or mutual catastrophe. But it is not war as that activity has been conceived for the last six thousand years.

In the same way that unleashing a pandemic using biological weapons would not be war—since it would spread to friend and foe alike—nuclear war, in which both sides would be catastrophically harmed, is not war.

We need a new category, a new concept for describing fighting done with weapons that are so destructive and hard to control that they damage both sides irreparably. Perhaps "mutual catastrophe." Perhaps some other phrase that captures the mutuality and lack of point.

NOT WEAPONS

Nuclear weapons are so clumsy (because of their size) that they are extremely hard to use in any sensible way. And this clumsiness is magnified by the radiation and fallout that the weapons produce. You could argue, in fact, that they aren't military weapons at all. And some prominent people have said just that.

For example, in July 1948, the Truman administration was debating who should have custody of nuclear weapons during peacetime. The national military establishment had formally requested that the bombs then in existence be turned over from the (civilian) Atomic Energy Commission to them. President Truman heard the arguments, thought the matter over for two days, and then rejected the request. According to one person present at the meetings (and who recorded the events in his diary), Truman explained his rationale this way:

> I don't think we ought to use this thing unless we absolutely have to. It is a terrible thing to order the use of something ... that is so terribly destructive, destructive beyond anything we have ever had. You have got to understand that this isn't a

military weapon.... It is used to wipe out women and children and unarmed people, and not for military uses. So we have got to treat this differently from rifles and cannon and ordinary things like that.[142]

Truman, the only man in existence who had ever ordered the use of nuclear weapons, seemed to have learned from that experience that a nuclear weapon "isn't a military weapon."[143]

Curiously, another leader who agreed with this assessment was the leader of the Soviet Union in the late 1950s and early 1960s, Nikita Khrushchev. His attitude toward the weapons was revealed in a private exchange of letters that occurred at the height of the Cuban Missile Crisis in October 1962. Premier Khrushchev and President John F. Kennedy exchanged private letters in which they urged on each other the importance of avoiding nuclear war. In one of these letters, Khrushchev tried to explain why he viewed the missiles that Russia had been sneaking into Cuba as defensive only:

> As a "military man," the President ought to understand that missiles alone, even a vast number of missiles of varying ranges and explosive power, could not be a means of attack. Missiles were nothing but a means of extermination. To attack, you needed troops. Unless it was backed up by troops, no missile—not even a missile carrying a hundred-megaton nuclear warhead—could be offensive.[144]

Khrushchev says nuclear weapons are "nothing but a means of extermination." In other words, using nuclear weapons could slaughter people and cause destruction; but that would not lead to victory in war. And this seems to sum up their utility pretty concisely: nuclear weapons are good at killing civilians but virtually useless in war.

WEAPONS EVOLUTION

The difficulties people have had in finding a situation where nuclear weapons were the right weapon for the job—the seven-plus decades that have gone by since their last use—are strong proof that they aren't very good weapons. But perhaps the most telling piece of evidence that their time has passed is the way all other weapons—the ones that are not nuclear weapons—have evolved over the last fifty or sixty years. There is a clear trend away from big weapons and toward small ones.

Over the last fifty years, the trend has clearly shifted toward accurate, intelligent, discriminating, small weapons. Smaller weapons have been used again and again, mass destruction weapons hardly at all. Smaller weapons are credited with playing a real part in victories in, for example, the Armenian/Ngorno Karabakh war and the war in Ukraine.[145] Blundering, massive weapons have no such victories to their credit, which makes sense. The point of war is to beat your adversary's military, not kill bystanders. Small weapons are inherently more flexible, more discriminating, more usable than big, blundering explosives. You can always use a really precise weapon. With a big weapon, you are constantly worrying that you'll accidentally destroy something you want or that you'll unintentionally harm your own soldiers.

Consider the Black Hornet—a tiny drone no bigger than half a cigar that hovers over the battlefield and peeks behind obstacles. It is small, elegant, cutting-edge. That is what the future looks like. The future belongs to miniaturized, nimble, swarming, adaptable, discriminating weapons—not big, blunt, blundering weapons.

Increasingly, nuclear weapons look like dinosaurs: giant, fearful creatures that once ruled the Earth but were eventually pushed aside by smarter, smaller competitors.

CONCLUSION

So, what sort of technology are nuclear weapons? If we had to sort all the different kinds of technology invented in the last one hundred years into one big list, where would they fall? Would they be at the top of the list as the most important technology ever invented? Would they be grouped with cell phones and jet travel in a special category for "most useful technology"? Would they be in a category called "most beneficial for humankind" that includes LED light bulbs and solar technology? It's a useful exercise to try to stand nuclear weapons technology next to other technologies and ask, "Just what sort of technology is this?"

It is easy to fall into the trap of imagining that nuclear weapons are useful because they can destroy things. This is exactly the trap that nuclear weapons advocates fell into (and remain stuck in). Nuclear weapons are really, really useful, they think, because they are really, really destructive. But nuclear weapons have a series of drawbacks that most people don't see because their eyes are drawn to the awesome mushroom cloud and the visions of cities turned to rubble. Hidden behind the smoke and images in our heads (which are, after all, merely symbolic) is a reality that includes substantial limits on the military applicability of nuclear weapons— limitations that only emerge when we look objectively and realistically.

Bernard Brodie, called the dean of nuclear strategists, talked about the difficulty of finding a real use for nuclear weapons:

> But the great and terrible apparatus which we must set up to fulfill our needs for basic deterrence and the state of readiness at which we have to maintain it create a condition of almost embarrassing availability of huge power. The problem of linking this power to a reasonable conception of utility has thus far proved a considerable strain.[146]

The arguments presented here are not meant to prove that nuclear weapons are useless in all possible scenarios. Clearly, there are some cases in which nuclear weapons might be useful. Attacking a tightly packed naval flotilla that was far out at sea, for example, might be a situation in which nuclear weapons would do the job. (Because the attack is on the water, it is unlikely that valuable resources, structures, or civilians would be nearby.) The point is not that nuclear weapons are completely useless. What this chapter argues is that an impartial assessment of nuclear weapons would have to admit that they have serious drawbacks. In fact, taking everything into account, it would not be an exaggeration to say that they are almost totally useless. They are, when you stop and think about them, about as desirable as lumps of coal. Nuclear weapons are weapons with hardly any military utility at all.

If nuclear weapons are nearly useless, is there a case for abandoning them? Should we admit that we made a mistake when we first evaluated their utility back in the Cold War, confess that we were wrong to build so many of them for so long, and simply declare that they are outmoded? Is there anything standing in the way of eliminating weapons that have very few practical uses? What other perspectives should we consider before making a declaration that they are obsolete?

[117] John Lewis Gaddis, *The Long Peace: Inquiries into the History of the Cold War* (New York: Oxford University Press, 1987) p. 119.

[118] Gaddis, *The Long Peace*, p. 125. Interestingly, Collins's conclusion mirrors the findings of a later study conducted during the Vietnam War, which found that U.S. bases and harbors made far better targets for bombing with nuclear weapons than the North Vietnamese forces in the field.

[119] Gaddis, *The Long Peace*, pp. 130–131.

[120] Bernard Fall, *Hell in a Very Small Place: The Siege of Dien Bien Phu* (Philadelphia: J. B. Lippincott Company, 1967), p. 425.

[121] F. Dyson, R. Gomer, S. Weinberg, and S.C. Wright, "Tactical Nuclear Weapons in Southeast Asia," Study S-266, Jason Division, DAHC 15-67C-0011, Washington DC, March 1967 (hereafter Dyson report). Declassified December 2002.

[122] Robert Green, *The Naked Nuclear Emperor: Debunking Nuclear Deterrence* (Christchurch, New Zealand: The Disarmament and Security Centre, 2000), p. 46. Insiders in Washington that I trust have told me that the rumor is that the study found that somewhere in the neighborhood of forty nuclear weapons would have been required. If you are curious about the details of such a calculation, see the unclassified paper by A. H. Nayyar and Zia Mian, "The Limited Military Utility of Pakistan's Battlefield Use of Nuclear Weapons in Response to Large Scale Indian Conventional Attack," Pakistan Security Research Unit, Brief Number 61 (November 2010).

[123] Gaddis, *The Long Peace*, p. 141.

[124] Elbridge Colby, Avner Cohen, William McCants, Bradley Morris, and William Rosenau, "The Israeli 'Nuclear Alert' of 1973: Deterrence and Signaling Crisis," The Center for Naval Analysis (April 2013), pp. 42–43.

[125] Avner Cohen, "The Last Taboo: Israel's Bomb Revisited," *Current History*, 104, no. 681 (April 2005), p. 172.

[126] United States Census Bureau, "QuickFacts New York City, New York," https://www.census.gov/quickfacts/fact/table/newyorkcitynewyork/PST045221 (accessed May 25, 2023).

[127] United States Census Bureau, "QuickFacts Nebraska," https://www.census.gov/quickfacts/NE (accessed May 25, 2023).

[128] This is similar to the frontage called for in U.S. military manuals in the 1980s. Battle Book (Center for Army Tactics) 86-(ST 100-3)-2202, 1 April 1986 Ft Leavenworth, KS.

[129] Since most modern divisions use a triangular formation of two regiments forward and one to the rear, the front is being covered by only two regiments, which means that roughly ten thousand soldiers are on the front lines. Half of that frontage (the area one regiment of five thousand soldiers would cover) would be 7.5 miles. Assuming that the front is no more than a mile deep, the square miles of the position would be 7.5 square miles, which, divided by five thousand, yields 666 soldiers per square mile.

[130] As I say in the text, these are very rough estimates and are not intended to be realistic estimates of the density of soldiers. They are only intended to give a sense of the difference between battlefields and cities.

[131] Gaddis, *The Long Peace*, p. 138.
[132] Gaddis, *The Long Peace*, p. 138.
[133] Gaddis, *The Long Peace*, p. 138.
[134] Quoted in Neal Rosendorf, "John Foster Dulles' Nuclear Schizophrenia," in *Cold War Statesmen Confront the Bomb: Nuclear Diplomacy Since 1945*, ed. John Lewis Gaddis, Philip H. Gordon, Ernest R. May, and Jonathan Rosenberg (Oxford: Oxford University Press, 2005, 1999), p. 81.
[135] See, for example, "Statement on Nuclear Weapons by International Generals and Admirals," signed by 60 retired generals and admirals from 17 countries, which reads, in part, "We military professionals, who have devoted our lives to the national security of our countries and our peoples, are convinced that the continuing existence of nuclear weapons in the armories of nuclear powers, and the ever present threat of acquisition of these weapons by others, constitute a peril to global peace and security and to the safety and survival of the people we are dedicated to protect." http://www.nuclearfiles.org/menu/key-issues/ethics/issues/military/statement-by-international-generals.htm (accessed May 25, 2023).
[136] "National Press Club Remarks, General Lee Butler, USAF (Retired), Wednesday, December 4, 1996, Washington, D.C.," https://nuclearweaponarchive.org/News/Butlpress.txt (accessed May 25, 2023).
[137] Freeman Dyson, Raymond Aron, and Joan Robinson, *Values at War: Selected Tanner Lectures on the Nuclear Crisis*, ed. Sterling M. McMurrin (Salt Lake City: University of Utah Press, 1983), pp. 18–19.
[138] Sidney D. Drell and Frank von Hippel, "Limited Nuclear War," *Scientific American*, 235, no. 5 (November 1976), p. 27.
[139] Drell and von Hippel, "Limited Nuclear War," p. 35.
[140] Arthur M. Schlesinger, Jr., *A Thousand Days: John F. Kennedy in the White House* (Greenwich, CT: Fawcett Publications, 1965), p. 363.
[141] Ralph E. Lapp, *Kill and Overkill: The Strategy of Annihilation* (New York: Basic Books, 1962), p. 7.
[142] Broscious, "Longing for International Control, Banking on American Superiority: Harry S. Truman's Approach to Nuclear Weapons," p. 30.
[143] I am aware that Truman did not truly order the use of the bombs dropped on Hiroshima and Nagasaki. A more accurate way of saying it would be that he "acquiesced in a plan brought to him to use nuclear weapons." Still, he was ultimately responsible for the outcome and appears to have felt this responsibility keenly. And I am also aware that in various public statements Truman spoke differently about nuclear weapons, never admitting that the bombing of Hiroshima and Nagasaki had been anything other than a military mission. I would argue that these private statements are more likely to be his actual views than things said on behalf of the nation in public.
[144] Quoted in Graham T. Allison, *Essence of Decision: Explaining the Cuban Missile Crisis* (Boston: Little, Brown and Company, 1971), p.221.
[145] See, especially, coverage of the Nagorno-Karabakh war, for example, Gustav Gressel, "Military lessons from Nagorno-Karabakh: Reasons for Europe to

worry," European Council on Foreign Relations, November 24, 2020, https://ecfr.eu/article/military-lessons-from-nagorno-karabakh-reason-for-europe-to-worry/ (accessed May 25, 2023); and Paul Iddon, "The Last Azerbaijan—Armenia War Changed How Small Nations Fight Modern Battles," *Forbes*, March 25, 2021, https://www.forbes.com/sites/pauliddon/2021/03/25/the-last-azerbaijan-armenia-war-redefined-how-small-nations-fight-modern-battles/?sh=1383ba157dd3 (accessed May 25, 2023).

[146] Bernard Brodie, *Strategy in the Missile Age* (Princeton, NJ: Princeton University Press, 1965), p. 273.

8. CRUELTY AND WAR

It is possible because cruelty doesn't win wars.

"War," the American Civil War general William Tecumseh Sherman famously said, "is Hell." When a southern woman berated him for the destruction his army caused, Sherman was unrepentant. He replied, "War is cruelty. There is no use trying to reform it; the crueler it is, the sooner it will be over." And he meant it. Shortly after saying this, he burned the twelfth largest city in the Confederacy (Atlanta) to the ground and then went on to lay waste to a swath of farms and homes from Georgia to South Carolina in what became known as his March to the Sea.

Sherman is not alone in believing that making war into Hell is the best way to make it short. John Arbuthnot (Jacky) Fisher, First Sea Lord of Great Britain in the years before World War I (and often considered after Lord Nelson the second most important figure in British naval history), shared Sherman's views. When he was invited to a peace conference at The Hague in 1899, Fisher didn't hesitate to voice his objections to humanitarian restraints on war:

> The humanizing of war? You might as well talk about humanizing Hell! The essence of war is violence! Moderation in war is imbecility!... I am not for war, I am for peace. That is why I am for a supreme Navy.... If you rub it in both at home and abroad that you are ready for instant war ... and intend to be first in and hit your enemy in the belly and kick him when he is down and boil your prisoners in oil (if you take any) ... and torture his women and children, then people will keep clear of you.[147]

IT IS POSSIBLE

Fisher, like Sherman, believed that cruelty had its uses. And this claim that cruelty can help to win wars is a key component in arguments about nuclear weapons. Some proponents assert that the cruelty that nuclear weapons inflict shortens (or even prevents) wars. Perhaps the most forthright exponent of this view was General Curtis LeMay. LeMay was the Air Force general who directed the U.S. bombing campaign against Japanese cities in World War II and ultimately became Air Force chief of staff. The historian Michael Dobbs described LeMay this way:

> In a single night, March 9–10, 1945, LeMay's B-29 bombers had incinerated sixteen square miles of downtown Tokyo, killing nearly a hundred thousand civilians. LeMay later acknowledged that he would probably have been tried as "a war criminal" had Japan won the war. He justified the carnage by arguing that it hastened the end of the war by breaking the will of the Japanese people.
>
> "All war is immoral," he explained. "If you let that bother you, you are not a good soldier."
>
> The object of war, LeMay believed, was to destroy the enemy as swiftly as possible. Strategic bombing was a crude weapon, almost by definition. The idea was to deliver a devastating knockout punch, without worrying too much about precisely what you were going to hit. In dealing with enemies like Nazi Germany, Imperial Japan, or Communist Russia, restraint was not only pointless, it was treasonous, in LeMay's view.[148]

LeMay believed nuclear weapons were decisive as long as you weren't squeamish.

According to Sherman, Fischer, and LeMay, there is no point in restraining violence in war. These views could hardly be considered politically correct today, and one could argue that enthusiasm for cruelty is a relic of a long-forgotten past. But it is surprising how often "civilized" societies have plunged into bloody and unrepentant war. So, perhaps it is only acknowledging reality to say that many wars involve widespread destruction and therefore it is only right to ask, "Can cruelty be a critical factor in victory?" Are we simply being delicate when we draw back from using gigantic weapons that kill civilians? Are Sherman, Fisher, and LeMay, in effect, right?

DESTRUCTION

Let's start by examining the role of destruction in wars. Obviously, destroying the means of making war—the factories that build the tanks, artillery shells, and other weapons soldiers use—weakens your adversary's ability to wage war. But hardly anyone would label blowing up military factories as cruelty. That is just waging war. To get cruelty, you usually have to have generalized destruction. To be cruel, you have to blow up houses and shopping centers and churches. In order to make war "Hell," you have to apply violence indiscriminately so that some non-soldiers get hurt. We're not exploring whether precision bombing of military targets is effective. We're asking, "Does general, indiscriminate destruction help win wars?" Does blowing up bars, daycare centers, and hospitals increase the likelihood that your adversary will surrender?

A careful examination of the history of war shows that aimless destruction almost never (and perhaps actually never) helps to secure victory. Two cases illustrate the role that wanton destruction plays in war.

NAPOLEON INVADES RUSSIA

Begin with the role destruction played in the invasion of Russia by Napoleon in 1812. After becoming Emperor of France in 1804, Napoleon changed the face of warfare by introducing massed armies. In the hundred years before, wars were generally fought by small, professional armies that often took trouble to avoid populated areas. After Napoleon's innovation of massed "citizen" armies, nineteenth-century war included pillaging the countryside, burning the cities, and fighting large-scale battles with increasingly high casualties. War raged across Europe for ten years, bringing millions of deaths, economic disruption, and widespread devastation.

In the summer of 1812, having conquered much of the continent, Napoleon marched into Russia at the head of a gigantic army.[149] This French and Allied force of 680,000 men looted and ransacked as it went, leaving destruction across a wide swath of Russia. Although exact figures are difficult to find, it is likely that a fairly large number of civilians were killed in this campaign. But despite losing key battles and surrendering hundreds of miles of territory, Russia's Tsar—Alexander I—refused to surrender. The Russian army mostly avoided direct engagements, continually retreated, and led the French deeper and deeper into Russia. Soon Napoleon's supply lines were dangerously long. To add to French supply difficulties, the Russians initiated a "scorched earth" policy: destroying crops and towns to deny them to the French. Even cities (Smolensk and Moscow) ended up being destroyed, although there is disagreement over who burned them.[150]

Napoleon's campaign left colossal havoc in its wake—some due to French soldiers wrecking and pillaging, some due to Russian preemptive destruction. The devastation of that war was on a scale with the greatest destruction ever visited on Russia up to that time. On French soil, however, there was an entirely different story. French fields, French livestock, French cities were almost entirely untouched. No campaigns were fought there; no invading armies stole and burned. The amount of destruction in Russia was enormous; the amount of destruction in France was negligible. If destruction wins wars, France won the war.

But of course, France didn't win the war. Despite the large-scale destruction of farms and villages and even cities in Russia, the Russians soundly defeated Napoleon, delivering what turned out to be the deciding blow in the Napoleonic Wars. Napoleon's armies, starved for supplies and unable to decisively engage the Russian army, were forced to retreat in disarray, losing more than ninety percent of their soldiers as they fled pell-mell through the harsh Russian winter. Even though the Russians suffered greater devastation, the French were the ones who were defeated.

DROPSHOT

The futility of destruction is not limited to the distant past. As a second example, consider the campaign that U.S. military planners imagined when they drew up the first war plan that included using nuclear weapons against Russia. Created in 1949 as a contingency, the plan was called "Dropshot"—and it called for the use of three hundred nuclear weapons on two hundred Russian cities and towns, along with additional conventional attacks.[151] This campaign of nuclear attacks was intended to destroy eighty-five percent of the Soviet Union's industrial capacity at one stroke.

What is striking about this plan, however, is that these military planners, most of whom would have been veterans of the campaigns of World War II, concluded that this devastating attack was unlikely to be decisive. After this initial bombardment with nuclear weapons—and the attendant destruction such an attack would surely cause—the plan called for a full U.S. mobilization and invasion of Russia. In other words, when these men, who had planned and carried out the most extensive bombing campaigns the world had ever known up to that time, imagined using three hundred nuclear weapons to devastate Russia, they concluded that such an attack would not finish the job. In their opinion, mere destruction on its own could not win such a war.[152] Russia, they concluded, would have to be invaded and conquered with land armies.

Their conclusion should not surprise us. After all, if you work your way back and forth across the span of human history, it is difficult (if not impossible) to find an instance in any war, from any era, on any continent where generalized destruction by itself won a war.

The rejoinder to Sherman—at least as far as destruction goes—is actually embedded in Sherman's own life experience. Despite the fact that, in the summer of 1864, Sherman captured Atlanta and burned it to the ground, the South did not surrender. Nor did the capture and destruction of Richmond—the capital of the Confederacy—the following winter force the South to give in. Even Sherman's March to the Sea—the deliberate use of destruction on a grand scale that left much of Georgia and South Carolina devastated—did not force the South to surrender. Only when Robert E. Lee's and Joseph E. Johnston's armies were surrounded and forced to capitulate did the war come to an end.

What history shows again and again is that nonspecific destruction, by itself, doesn't win wars. Cruelty in the form of destruction doesn't seem to break people's will—or if it does, "breaking the will of the people" is apparently far less important than "breaking the will of the military." No nation, facing a real prospect of being conquered and still in possession of substantial military power, has ever surrendered because houses and farms were burned and destroyed.

KILLING CIVILIANS

But perhaps Sherman was simply not being cruel enough. Perhaps for cruelty to work, it has to include not just indiscriminate destruction but also generalized killing—killing of civilians.

One way to examine this notion is to look at one of the cruelest and most horrifying military campaigns in history: the savage war Genghis Khan waged from 1219 to 1220 to conquer the Khwarazmian Empire. The Khwarazmian Empire encompassed much of what is now modern-day Iran, Afghanistan, Turkmenistan, and Uzbekistan and also included parts of what are now Kazakhstan, Kyrgyzstan, and Tajikistan. It sat squarely astride the trade routes from east to west, and contemporary accounts paint a picture of a land "fertile, rich and phenomenally flourishing."[153] Fed by elaborate irrigation systems, its cities were surrounded by flowering plains of dates, oranges, melons, and other delicacies.[154] A number of those thriving, long-ago cities are estimated to have held as many as a million people.

The Mongols swept out of the east in late 1219 and destroyed city after city. Some of these cities were burned; some were torn stone from stone. One—Urganch, the capital—was inundated. The Mongols opened the dams on the Amu Darya and allowed the city to be completely covered by water. The destruction of each city was generally accompanied by horrifying scenes of slaughter. At Merv, for example, the Mongol commander, Genghis Khan's son, is said to have had a golden throne set on the plain outside the city so he could watch at leisure while something like six hundred thousand men, women, and children were slaughtered.[155] Only four hundred "artisans" were spared.[156] Although the numbers may be suspect, again and again, reports of unimaginable killing come down to us from multiple reliable sources.

And the scale of the slaughter is not the only aspect of the Mongols' cruelty. In other places, the Mongols are supposed to have piled the heads of the city's populous in pyramids, or forced civilians to run around the city while Mongol soldiers fired arrows from the walls, using them for target practice. There is a reason that Genghis Khan's name is still a byword for cruelty.

More than a dozen cities were attacked, ransacked, leveled, and their populations wiped out. Historian Louis Dupree summed up the Mongols' impact by saying, "Genghis Khan was the atom bomb of his day; and western Asia still bears the scars, still suffer[s] from the economic impact."[157] The war that the Mongols waged devastated civilization in that part of the world—a blow from which it has yet to recover.

But the unimaginable slaughter did not lead to the surrender of the Khwarazmian Empire. Killing so many civilians did not bring the war to an end. We know this because even two years later, despite the destruction of more than a dozen cities and the deaths of possibly four million civilians, and despite scenes of frightful cruelty, Genghis Kahn was still having to fight. The war did not end until the last army—commanded by the son of the late Khwarazmian Shah—was defeated on the banks of the Indus in 1221. Cruelty did not lead to victory. Defeating the remaining Khwarazmian armies was the necessary step.

Of course, you could argue that the failure to coerce surrender in this case resulted from not enough civilians having been killed. Since nuclear war would kill civilians in unprecedented numbers, perhaps the explanation here is that Genghis Khan's soldiers didn't murder enough residents of Khwarazmian cities to make the tactic work. Maybe the Great Khan wasn't cruel enough somehow.

There is, however, a historical case that addresses this kind of speculation.[158]

An extreme and appalling example of the extent to which civilians can die without stopping a war can be found in the Paraguayan War of 1864–1870. An unyielding leader, Francisco Solano López, led Paraguay in a war against the combined forces of Brazil, Argentina, and Uruguay. Exact figures cannot now be reconstructed, but credible assessments put the civilian losses at something like sixty percent of Paraguay's total population. This means that during the course of the war, as civilian losses mounted, López and the Paraguayan military—aware of the large numbers of civilians dying—repeatedly chose to fight on. When thirty percent of the civilian population had died, they did not say to themselves, "Well, too many civilians have died. Now we must stop fighting." When forty percent of the civilian population had died, they did not stop to reconsider and say, "Well, now too many civilians have died." And so on. They were not, apparently, influenced at all by the cruelty being inflicted on Paraguay's civilians. Only when López himself was killed, and the last military forces disintegrated, did the war finally come to an end.

Sixty percent casualties is roughly in line with some of the worst predictions in all-out nuclear war scenarios. If some level of civilian losses could cause a society to disintegrate or reliably force capitulation, one would imagine that having sixty percent of the population killed would be sufficient. But apparently it wasn't.

The history of war is quite clear: killing civilians does not lead to surrender. Perhaps you can find some marginal cases where killing civilians affected a leader's decision to cease fighting—some war of choice where the stakes were low (although I have yet to find such a case). But in any war where the survival of the state was at risk, the deaths of civilians have never led to surrender. In fifteen years of trying, doing my best to canvass all parts of the globe and all eras of history, I have never found an instance of a leader saying, "Now we must surrender because the cruelty that civilians are suffering is too great." And even if an exceptional case (or two) exists, that only highlights the remarkable hardness of the rule: killing civilians doesn't win wars.

THE AIM OF WAR

Those who argue that attacking civilians is an effective means of waging war have forgotten what war is about. War is not about "killing people and breaking things." Russia's leaders during the Napoleonic wars understood that while suffering destruction was painful, it was not militarily decisive. They could despoil their own crops, burn their own citizens' farms, allow peasants to starve, and still win the war. Russia's leaders wisely saw that what was at the heart of war is the soldier. Beat the soldier or cause them to give up and you win the war. In wartime, state power is dependent not on civilians but on soldiers.

Civilians in war are largely beside the point. When Wyatt Earp and Doc Holliday fought their famous gun battle at the OK Corral, it was not the side that shot the most bystanders that won. It was the side that shot the most gunslingers. The same is true of war. Killing bystanders is largely irrelevant. Killing soldiers (or forcing their surrender) is the key.

Secretary of Defense Robert McNamara made this point in a famous speech talking about U.S. policy for fighting a nuclear war: "[I]n the event of a nuclear war, [the objective] should be the destruction of the enemy's military forces, not of his civil population."[159]

We sometimes forget how little practical value cruelty has because our eyes are drawn to the horror of dead civilians and burning cities. The scar left by these gruesome events can remain livid for centuries. But the fact that this sort of horror is seared into collective memory does not change the rules of war. Only when militaries are defeated do wars end.

It is difficult to argue from the facts that killing civilians or creating random destruction—cruelty, in other words—wins or even shortens wars.

* * *

People who argue in favor of keeping nuclear weapons sometimes talk with "tough guy" bravado about the ability of nuclear weapons to create suffering. They use phrases like "War is Hell" or quote military strategists on the importance of the power to inflict pain. They sometimes argue, as Thomas Schelling does, that nuclear weapons have real military utility that can be converted into effective coercion. But there is little evidence that such military utility exists. Wars are not won by killing civilians. In a fight to the finish, civilians simply don't matter enough to make a difference.

Bernard Brodie, after examining the bombing campaigns of World War II, wrote that "the effects of bombing on civilian morale were certainly not trivial, but it seems clear that the lowered morale resulting from bombing did not importantly affect military operations or the outcome of the war." Attacks on enemy morale, based on the evidence of World War II, were "a waste of bombs."[160]

The argument that nuclear weapons can be useful if we are just steely-eyed enough is one important strand of the rationalization for keeping nuclear weapons. Schelling's pain arguments are taught in graduate schools and repeated in the halls of government. It is emotionally pleasant, for a certain kind of person, to imagine that they are tougher than the rest of us. There is, too, a certain appeal in the thought that you see harsh realities that others are too weak to face. But confusing pleasant emotions or imagined toughness for reality is not the sort of mistake a genuine realist makes. It is not realism to mistake cruelty for military utility.

CRUELTY AND WAR

If we don't need to keep nuclear weapons because they are cruel—and we've already seen that they have little military utility—are we then ready to declare them obsolete? Or is there another rationale for keeping these blundering, dangerous weapons?

[147] Massie, *Dreadnought*, p. 431.
[148] Michael Dobbs, *One Minute to Midnight: Kennedy, Khrushchev, and Castro on the Brink of Nuclear War* (New York: Alfred A. Knopf, 2008), pp. 96–97.
[149] For an account of the invasion, see David G. Chandler, *The Campaigns of Napoleon* (New York: Macmillan, 1966).
[150] There are still arguments over whether the burning of Moscow was intentionally done by the Russians, was an accident, or was done intentionally by the French.
[151] David Alan Rosenberg, "The Origins of Overkill: Nuclear Weapons and American Strategy 1945-1960," *International Security*, Vol. 7, No. 4 (Spring, 1983), pp. 3-71.
[152] "This [1950] was still not a period when it was taken for granted that all-out war meant the destruction of whole societies. The Harmon and Hull reports of 1949 and 1950 had made it clear that the initial 'atomic blitz' would not be counted on to destroy the war-making power of the Soviet Union.... It was thus taken for granted in the early 1950s that a third world war would be long. In the first few weeks of the war, the United States would be swept off the continent of Europe, at least up to the Pyrenees. America would then begin to mobilize its resources and mount a sustained bombing campaign with atomic bombs and aircraft produced after the outbreak of the war. The Soviets, who now had the great resources of Western Europe to draw on, would at the same time be conducting their own air offensive against the United States and its bases and allies overseas. This would be a war of endurance." Trachtenberg, *History and Strategy*, pp. 119–120.
[153] David Morgan, *The Mongols* (Cambridge, MA: Blackwell, 1986), p. 82.
[154] It is perhaps hard to imagine that this part of the world flourishing to this extent, but consider how irrigation transformed the San Joaquin valley in California from desert to garden.
[155] For more on how to think about the difficult-to-believe scale of the casualties, see the balanced discussion in Morgan, *The Mongols*, pp. 73–81. Morgan concludes that a careful reading of the sources makes these numbers seem credible but definitive answers must await future archeological excavation of the cities.
[156] J. J. Saunders, *The History of the Mongol Conquests* (London: Routledge & Kegan Paul, 1971), p. 60.
[157] Louis Dupree, *Afghanistan* (Princeton, NJ: Princeton University Press, 1973), p. 316.
[158] And let's be clear, without any historical evidence to prove that killing civilians wins wars, the claim that wars can be won if you apply enough cruelty is no more than speculation.
[159] Quoted in Lapp, *Kill and Overkill*, p. 87.
[160] Brodie, *Strategy in the Missile Age*, p. 103.

9. SYMBOLS

It is possible because nuclear weapons are more symbols than weapons, and symbols make poor defenses.

If nuclear weapons are not very good weapons, how do we explain the importance they have in the world? Why do they play such an important role in some interactions between countries? The answer has to do with the peculiar power of symbols and the strange dual role that nuclear weapons play.

*   *   *

The ancient Romans were a harsh and self-disciplined people. Their military achievements were not the result of exploiting superior technology, improved battle tactics, or the genius of a single leader. Their military dominance was built on a particular attitude to war. Perhaps this attitude was best summed up by Winston Churchill (whose education undoubtedly included a healthy dose of Roman literature and history) when he gave a speech at his old school, Harrow, in October 1941. He said: "Never give in, never give in, never, never, never, never—in nothing, great or small, large or petty—never give in except to convictions of honour and good sense. Never yield to force; never yield to the apparently overwhelming might of the enemy." The Romans put this kind of relentless determination at the center of their culture. It characterized their politics, their society, and, especially, the way they made war.

For example, when tensions between Rome and Tarantum, the chief city on the heel of the boot of Italy, led to war in 282 BC, the leaders of Tarantum knew that Rome would send an army to attack them. Fearful,

IT IS POSSIBLE

they called on King Pyrrhus of Epirus (across the Adriatic in Greece) to defend them, and Pyrrhus, hoping to get a foothold in Italy, not only sent an impressive army but came himself. The Roman force raised to retaliate against Tarantum and eject Pyrrhus from Italy was a well-equipped and fully manned army under the consul Publius Valerius Laevinus. In the event, Pyrrhus soundly defeated the Romans at the battle of Heraclea in 280 BC, inflicting as many as fifteen thousand casualties. Pyrrhus undoubtedly looked out over the battlefield with satisfaction and triumph. But the next spring the Romans were back. They had raised a fresh army, trained and equipped it, and brought it (led by a new consul) to oppose Pyrrhus. Another battle was fought, and again, the king's army was victorious, this time inflicting six thousand casualties. Pyrrhus again looked out over a battlefield strewn with the bodies of his enemies. But he couldn't help noticing how many of his own soldiers were also dead on the field. He is supposed to have remarked, "If we are victorious in one more such battle with the Romans, we shall be utterly ruined."

And of course, the Romans then raised still another army, calling up thousands of fresh recruits. After spending a year campaigning in Sicily, Pyrrhus returned to the mainland to find himself vastly outnumbered by this new force. Pyrrhus had won every battle. But in the end, he had to withdraw from Italy and leave Tarantum to its fate. Pyrrhus had learned (the hard way) the lesson that so many of Rome's enemies would learn: Rome was relentless. Unless you completely wiped out the entire Roman people, they would always return to the fight.

The Romans, in a way, showed their enemies a kind of grudging respect. They imagined that their adversaries, given the chance, would be as tough and relentless as they themselves were. If they did not destroy their enemies utterly, they reasoned, those enemies would do to them what they did to others: come back to fight again and again until they won or were utterly destroyed. Only the harshest penalties, the Romans felt, could ever dissuade a determined adversary from continually coming back to fight again. As a result, Roman punishments, both in international relations and in the administration of civil justice, were unusually severe. In the Roman mind, harshness was necessary because of the human capacity for determination.

As a result, subject peoples who broke Roman laws were often punished with brutal severity, at least by our standards. One of the cruelest of these punishments involved nailing two long pieces of wood together and then using large spikes driven through the criminal's arms and legs to pin him to this roughly T-shaped structure. Usually, people who were punished in this way died slowly of exposure and loss of blood. After the vast slave revolt of 73 BC—a revolt depicted in the classic Hollywood movie Spartacus—the Romans took the six thousand slaves captured after the fighting ended and nailed them up to these man-made trees for miles along the Appian Way. It is said that the slaves' bodies—pinned to these wooden structures—lined the road from Rome all the way to Capua, a distance of 120 miles.

We are familiar with this Roman practice because it was the punishment meted out to a rebellious religious leader in Judea in the nineteenth year of the reign of the Emperor Tiberius. It was a time of exceptional turbulence in that Roman province, and the Romans hoped, apparently, that they could stamp out his new religion and quell the unrest by murdering the leader of this strange sect. But the followers of this rebellious leader reacted in a remarkable and unexpected way to Roman harshness. Rather than allowing this harrowing act of cruelty to dictate the meaning of their leader's death, rather than feeling shame or horror, rather than turning their backs on the memory, they focused on it and absorbed that memorable punishment into the fabric of their religion. Rather than forgetting the injustice and humiliation of their leader's end, they took the structure that had been used to kill him and made it into a sort of badge of their faith. The two pieces of wood—one longer piece placed vertically and a second, shorter piece fixed across it—were taken as a remembrance of their leader and all that he had stood for. Rather than hiding it away and reviling it as the instrument of their leader's cruel end, they inverted the meaning of this wooden structure and made it into the central symbol of their faith.

Today, of all the many symbols that human beings use to bring meaning to their lives, the Christian cross is probably the most recognized symbol in the world. What was intended as a harsh warning of Roman relentlessness and determination became instead an inspiration of Christian faith. As of 2020, some 2.5 billion people, a little less than a third of the world's population, looked to crosses as the paramount symbol of their religious belief. Crosses hang on walls around the world. They prompt people to recall their commitment to the precepts taught by that rebellious religious leader long ago. Millions of crosses hang around the necks of Christians—touched again and again during the course of a lifetime—as a talisman, a token, a physical manifestation of inner spirituality, and a set of beliefs. And inexplicably, those pieces of wood, those small pieces of jewelry, that characteristic shape, bring meaning to the lives of those who see and touch them.

Symbols are a strange and powerful part of human experience. They give life meaning, and meaning is, in some ways, the most important characteristic a life can have. People sometimes say, "I don't care if my life is short, as long as it means something." But despite the apparent importance of meaning in our lives, we don't really understand symbols. How can a thing, an object create meaning in our lives?

It is easy enough to imagine an event giving meaning. That makes sense. You meet Martin Luther King Jr. and shake his hand. For the rest of your life, you remember that moment and feel what seem like reverberations of meaning emanating from the encounter. Funerals, rallies, sporting events, raging arguments, quiet walks with those we love, deathbed farewell words—of course, these kinds of events can generate meaning. But how does an object generate meaning? A physical object—which can be made from entirely ordinary materials—somehow finds itself having the power to stand for something else while at the same time evoking feelings and values. It, as it were, embodies meaning, as if the meaning were some invisible, magic something that could be gathered like a breath and blown into that physical object and then somehow become a reminder of things that can't be expressed. The American flag, for example, reminds some people of everything that the country means to them: the heroes, institutions, accomplishments, landscapes, patriotic songs, traditions, foods, wars, scientific achievements—all these are somehow embodied in cloth of three colors sewn in a certain pattern. But how is this alchemy achieved? We don't know. We don't know how meaning works, where it comes from, or how it can be most accurately represented.

What we do know is that nuclear weapons have become symbols.

BECOMING SYMBOLS

Not all weapons become symbols. Few people would say that a hand grenade, for instance, has a symbolic meaning. There is no romance to a hand grenade; it doesn't resonate with a special feeling. It is more like a tool you use without giving it a second thought.

But it was, in some ways, almost inevitable that nuclear weapons should have become symbols—that they would seem to carry a larger freight of meaning than other types of weapons. From the beginning, they weren't just bigger bombs. The brilliant flash of light, the all-pervading roar, the immense shockwave, and the towering cloud—everything about a nuclear explosion fills us with awe.

The reactions of the scientists and military men who observed the first test, searching for words big and profound enough, illustrate how meaningful the weapons seemed to be to them. Even for people who were not there, that first detonation reverberated with religious, biblical, and godlike overtones. It wasn't long before people began to refer to these weapons as "the Bomb"—in the singular, with a capital "B"—the way they referred to the one God with a capital "G." President Truman, after hearing about the successful test and approving the order to use the bomb against Japan, summed up his feelings in a long diary entry on July 25, 1945, saying,

"We have discovered the most terrible bomb in the history of the world. It may be the fire destruction prophesied in the Euphrates Valley era, after Noah and the fabulous ark."[161]

Instead of thinking analytically, the small circle of scientists and government officials who knew about the bomb felt impelled to explain the test as a representation of a larger, more portentous religious or mythical force.

In many ways, nuclear weapons were perfect objects to use as symbols. They had several crucial elements needed for a symbol. They were awe-inspiring. They had towering clouds of smoke; they had a deep rumbling roar; they had a blinding flash of light. They immediately reminded people of godlike power. When they were used in war, they seemed to create a miracle: forcing a recalcitrant foe to surrender in just four days.

But it was also what they lacked that made their exponential growth as symbols possible. What they lacked was real-world experience. The absence of regular interaction with nuclear weapons is one of the key factors in allowing their symbolic value to expand and spread into areas where no weapon had had meaning before. Nuclear weapons aren't like every day plates, bowls, cups, and saucers. Used over and over, chipped and worn with constant handling, these familiar objects eventually lose any aura of specialness. Nuclear weapons are, rather, the good china, kept reverently locked away behind glass. They are never brought out except on the rarest of occasions, like holy relics of inestimable value. Objects treated with such care and respect must have awesome powers indeed.

And more importantly, nuclear weapons have no long record of success or failure, no uncomfortable body of evidence to restrain our imaginations from running wild. Nuclear weapons were used once in 1945, three days apart, against two cities, and since then, they have not been exploded during war again. In this vacuum, their symbolic power has run rampant, growing and expanding with amazing speed. If you are going to invent a god, you want the manifestations of that god—the times when they are actually present in the world—to be both intensely impressive and extremely rare. Familiarity, as the saying goes, breeds contempt. Nuclear weapons have avoided this fate admirably by only being used once. They make perfect symbols. We are free to assign almost whatever meaning we want to them.

TOO MODERN FOR SYMBOLS

Of course, some people will object to this description of nuclear weapons as symbols. "Symbols" they will say, "may have been powerful in the ancient world. And, of course, the symbols that got embedded in religion have tended to stay there and maintain their importance. But we're modern. We've lived through the Enlightenment. We don't get caught up in all that prehistoric mumbo-jumbo of symbolism that impressed primitive people in the ancient world. What's impressive about nuclear weapons is what they can actually do. I'm only interested in the real weapon, not made-up imaginings." And you can see that this would be a very common reaction to talk of nuclear weapons and symbols. Most people, after all, think of themselves as rational, practical individuals. At least, most Americans do. We pride ourselves in dealing in dollars and cents, not dreams. Most of us would bridle at the suggestion that symbols have much power or even much of a role in our lives.

But although they sometimes escape our notice, almost everyone's life is filled with symbols. Symbols fuel our emotions and influence our thinking. They shape the way we frame discussions and motivate some of our most important actions. Most people, when they look at a rose, don't just see a specific variety of plant and nothing more. Almost everyone sees a rose as a symbol of love. Some look at a varsity letter jacket from a university and don't see an article of clothing useful on a cool day. They see prestige and accomplishment and social standing. Human beings use symbols all the time, and those symbols, in turn, enrich the meaning of their lives.

Ironically, even the phrase I used before—"dollars and cents"—which is meant to connote people whose feet are planted firmly on the ground, illustrates perfectly how deeply our lives are shot through with symbols. Money, one of the most important things in our lives, is entirely symbolic. The currency in your pocket is nothing more than pieces of paper with pictures and numbers on it (or even more symbolic, simply numbers on the screen of your phone). We agree that these symbols will represent so many hours of work, so much market value, or whatever. But that value is in no sense intrinsic to the pieces of paper, the plastic card, or the numbers on a screen. We handle money every day and think of it as the most practical and down-to-earth commodity. In fact, it is the best proof there is that human beings constantly make and use symbols.

So, it is not surprising that nuclear weapons became symbols. Symbols can be manufactured out of anything. We turn roses, flags, crosses, diamonds, and pieces of paper with pictures of presidents on them into symbols. Sometimes we turn military weapons, like battleships, into symbols. These symbols vibrate with a meaning that goes beyond their practical value. And, in some cases, it is the symbolic value, rather than their ordinary utility, that makes them important.

So, nuclear weapons are symbols. Let's turn now for a bit to examine how we interact with those symbols.

ORACLE ON THE SUBWAY

I once met a stranger on the subway in Brooklyn, standing on a crowded No. 5 train. He was about thirty or so, and his clothes were worn and dirty. Not the kind of dirtiness that comes from working hard for a day, but the kind of ground-in dirt, frayed edges, and smell that comes from a person who has been sleeping on the ground in these same clothes—and probably hasn't changed them—for many days. He was carrying a dirty bed pillow under his arm, which was incongruous at four o'clock in the afternoon on the No. 5. Even though it was warm, his sweatshirt hood was over his head.

Looking back, I'm sorry to say that what probably flashed through my mind when I first glanced at him was something like "Dirty clothes—smell—huh, a pillow. Must be homeless." And I went back to my reading. Homelessness was not that extraordinary in New York at the time. But my older self looks back with a certain disappointment at my unwillingness to engage in a human way, my dismissiveness.

A moment later I realized he was talking directly to me. "Does it walk, baby?" he said, his eyes forcing the question on mine. I looked back uncomprehendingly. He said again, "Does it walk?" and pointed to the copy of the ancient Greek tragedy, *The Libation Bearers* by Aeschylus, in my hand. And then he launched—using a sort of singsong voice—into a paragraph or two of what I later realized was loose quotation from a long section of the central part of the play, interspersed with his own comments on the meaning of the work. It was hard to follow. Some of it, I'm pretty sure, was gibberish. But some of it sounded freakishly profound. I felt the kind of shock and disassociation you feel when you wake out of a sound sleep and for a moment have no idea what room, house, or even country you are in. I had been lost in a scholarly, intellectual world—distant, safe, a place of judicious and unemotional contemplation—that had suddenly been invaded by a strange, compelling, incomprehensible oracle with mystery and prophecy tumbling out of his mouth.

The hairs on the back of my neck stood on end. I stared at him, trying hard to follow what he was saying and feeling the creeping awe and fear one always feels in the presence of the profound. We stood and he talked for what seemed like a long time but which was probably only five minutes. For a brief moment, I wasn't sure if I was still in the same reality—if anything still made sense. Then the train pulled into Nevins Street station, he looked me in the eye like a teacher with one last profound message to impart, and said deliberately, "Does it walk?" and got off the train.

When I read certain experts and even government officials talking about nuclear weapons, I'm often carried back to the clash between meaning and reality that I had on the No. 5 train. I feel again the sense of dislocation between the powerful and profound meaning that seemed to lie behind his words and the reality of a man with a strong smell holding a grimy pillow. A number of people—a surprisingly large number of people—use a curious kind of incoherent double-talk when they discuss nuclear weapons, almost as if they were suffering from some sort of mental defect. I'm not talking about the jargon or the arcane theories. I'm talking about what seems like a divide in their thinking and a companion tendency to (apparently) contradict themselves.

A quite famous example of that sort of disassociation comes from Mao Zedong, the leader of China throughout much of the last half of the twentieth century. Mao famously called nuclear weapons "a paper tiger" (in other words, a fake danger and not a real threat). It was a widely quoted dismissal of what many think of as the most powerful weapons in the world. But then, on a different occasion, Mao called them "a real tiger."[162] This kind of direct self-contradiction is disorienting. Real tiger or paper tiger—which is it?

And Mao wasn't alone. Take John Foster Dulles, who was U.S. secretary of state (and Mao's adversary) under President Eisenhower. Dulles had an equally confusing duality when he talked about nuclear weapons. He had what one historian politely calls "oscillating thinking" about them. "Dulles experienced swings back and forth between anxiety over the atomic bomb's terrible potential for destruction ... [and] eagerness to use the bomb as a diplomatic, and occasionally military, club."[163] Dulles, who in the 1950s famously made some of the most frightening threats with nuclear weapons ever issued, seems as if he couldn't make up his mind about whether they were horrifyingly destructive or amazingly useful.

And it's not just government officials who talk this way. Lots of people who think about nuclear weapons display a similar willingness to apparently contradict themselves. John Mueller, for example, a scholar who is well known for doubting the importance of nuclear weapons, has said, "While nuclear weapons may have substantially influenced political rhetoric, public discourse, and defense budgets and planning, it is not at all clear that they have had a significant impact on the history of world affairs since World War II."[164] This creates consternation among some of Mueller's readers. If nuclear weapons had so much influence, how can they have had no impact on world affairs?

This tendency toward contradiction is reflected in the larger debate as well. The people who think we should keep nuclear weapons, for example, regularly say that no one really intends to use nuclear weapons and, at the same time, that we use nuclear weapons "every day." It's as if they are arguing with themselves—first taking one side of the question, then the opposite one. You might almost be tempted to say that there is a touch of disordered thinking in these people, which is kind of troubling because these disordered thinkers are dealing with nuclear weapons. Is it possible that the responsibilities of thinking about the end of civilization or spending so much time contemplating the horrors of nuclear war have somehow shaken loose the sense in these people's heads? Does this frightening subject somehow destabilize the minds of the experts and government officials who spend years focused on it?

Fortunately, it is not a mental defect that creates these strange utterances. This behavior has a more prosaic cause. It is, in fact, a flaw that lies at the heart of the nuclear weapons debate, a flaw that—once you see it—explains many of the contradictions and peculiarities that surface in what people say. The source of these apparent contradictions and the secret to understanding much of the confusion surrounding these weapons is symbolism. In order to understand what is going on in the debate, you have to remember that nuclear weapons are two things: They are both weapons and symbols. They are not, like the dinner fork on the table, merely one thing. They are, so to speak, the thing-in-itself (*Ding an sich*) and at the same time the ghostly image of the thing that glows in our minds—the symbol, both together at the same time. Our behavior toward them can't fully be described without paying attention to both the weapon and the symbol.

Nuclear weapons are the practical weapons whose destructiveness has been carefully studied, the practical weapons that so few people really want to use. But they are also—at the same time—the symbolic weapons that are used for threatening, strengthening alliances, and status. The reason people talk in two different ways about nuclear weapons is that they are flipping back and forth between the two natures of these weapons. Unconsciously they are talking about the blundering weapon and the mighty symbol of power at the same time without signaling that they are jumping from one to the other, from realism to symbolism.

One reason this seems like mental disorder is that weapons and symbols are completely different types of things. They are not like pepper and paprika, which both share the same parent class (spices), which both share all the characteristics that are typical of that class, and which are only different because of their tastes. Weapons and symbols come from entirely unlike classes of things. They are like waves and particles—one a form of motion spread over an entire phenomenon, the other a speck of something in a particular place. The easiest way to see how stark these differences are is to examine the different ways that they are used.[165]

IT IS POSSIBLE

First, let's think about them in their incarnation as weapons. Weapons belong to the larger class of tools, which are implements we use to enhance our ability to directly change the physical world. A trowel moves dirt. A multi-tool bladed knife cuts a line, files a rough place, or opens a beer bottle. A tank fires projectiles that explode and violently rearrange earth, buildings, vehicles, or people's bodies. These are tools, and their chief characteristic is that they extend our ability to change the material world in some measurable way.

Contrast this with their incarnation as symbols. If I had an oblong piece of plywood hanging on my wall painted neon orange and you said, "What's this?" and I said, "It's a tool," you would probably scratch your head and ask, "What does it do?" Because tools are characteristically things that do something.

If I responded, "Oh, nothing. I've had it since childhood, and it reminds me of our summer home in Canada," there would probably be a slight pause. And then you would think to yourself, "An orange oblong piece of wood that hangs on the wall that doesn't do anything is not a tool." And you'd be right. My orange oblong is a memento, a keepsake, or, perhaps, a symbol.

Unlike tools, symbols do not directly change the world. They can only affect the world through someone. A soldier sees the flag, is inspired, fights heroically, and turns the tide of the battle. A woman sees a rose on a table, is reminded of all her feelings of love toward her husband, and the two are reconciled. The symbol changes the outcome but only through the agency of a person. Symbols have to mean something before they can impact the world. But only human beings can construct that meaning. You could think of symbols as the fuel that powers the little engine of our hearts. Symbols change how people feel; they motivate people, and only then does action follow.

Another way to think about this is to note that tools can be used mindlessly. A person numb with grief or suffering from dementia can still dig with a shovel—their "muscle memory" allows them to complete the task even though their conscious brains are not engaged. But someone who is not thinking at all is not affected by a symbol. Symbols only work when some as-yet-undefined process takes place in our heads—perhaps a strange amalgam of thought, imagination, inspiration, and other emotional reactions. In other words, something has to happen inside of us before symbols can "work."

There is a natural contrast in our minds between symbols and tools. They are used, after all, for quite different purposes. But the contrast is especially stark when the tools and symbols are nuclear weapons.

Detonating a large number of nuclear weapons would cause catastrophic harm to the people where the weapons exploded and to others downwind. Used as tools of war, nuclear weapons represent colossal danger. Paradoxically, nuclear weapons advocates claim that their symbolic use is enormously beneficial. They say that nuclear deterrence keeps us safe, solidifies our alliances, makes our nation great, strengthens the world order, and makes prosperity and peace possible. Nuclear weapons, therefore, appear in the current debate like a shape-shifting god. They move rapidly and unexpectedly between two entirely different forms. In one role (their weapon role), they are like a terrifying primeval monster-god—giant, remorseless, and destructive. In their other role (as symbols), they are like Demeter—the benevolent goddess of the harvest whose life-giving abundance is responsible for much of the good in the world. This is confusing: Demeter–Monster-god; Monster-god–Demeter. Which is it?

It is this dual nature of nuclear weapons—which often passes unnoticed—that is one of the main reasons the debate about them is so difficult to comprehend. So, for instance, an expert might believe that the ordinary object (the weapon) was dangerous and horrific but that the symbol was useful and awe-inspiring. But if he decries the horror of nuclear weapons in one breath only to praise the power of nuclear deterrence with the next, he will deeply confuse his audience. Because experts and officials so often fail to acknowledge the dual nature of nuclear weapons and generally don't identify which version of the weapons they're talking about (the symbol or the weapon), they sometimes create a sense of alarm and confusion in their listeners—similar to the feeling I had when I met the prophet on the subway with a pillow under his arm.

SYMBOLS FIRST

It's worth pointing out that while nuclear weapons have two roles, these two roles are not equal. When you compare their symbolic uses to their actual uses, they are clearly symbols first and weapons of war only a distant second. If we value nuclear weapons, it is primarily because we are attached to them as symbols, not because they are particularly good weapons.

The first and most persuasive piece of evidence supporting this conclusion is that they don't get used as weapons. They aren't actually detonated. Across the seventy-five-plus years since Nagasaki, they haven't physically interacted with the world even once in wartime. And this nonuse is not because of a lack of opportunities. The nuclear-armed states have fought wars again and again during those many decades. The United States fought large wars in Vietnam, Korea, Iraq, and Afghanistan, not to mention quite a few smaller conflicts. The Soviets fought in Afghanistan for ten years, in Chechnya, and now in Ukraine. China fought a border war with Vietnam in 1979. The British fought the Argentines in 1982. And so on. These conflicts were all occasions when other weapons of war—tanks, jets, AK-47s, cruisers, soldiers, and so on—were used. But not nuclear weapons. These were conflicts where soldiers died, where important national interests were at stake, where sacrifices were made. Each of these conflicts were opportunities for nuclear-armed states to use these awesome, powerful weapons to prevent their own soldiers from dying, and to ensure their interests were secure. But the nuclear-armed states did not use their nuclear weapons.

The same could not be said of their symbolic role. If anything, their symbolic role has expanded over the years. We use them symbolically to buttress our treaties. We use them symbolically to deter our adversaries. In fact, nuclear weapons advocates argue that we use them as deterrents "every day." They argue, in other words, that, although with only one exception we never use them as weapons, we use nuclear weapons constantly as symbols.

Obviously, they could be detonated. They could materially interact with the world. It's not as if they are an oblong piece of plywood hung on the wall that literally can't do anything. Nuclear weapons are mechanisms that have the potential to radically transform the material world. But to date, we've chosen not to detonate them. Instead, we talk and think about them, we allow them to resonate. In other words, we choose to interact with them using our imaginations and emotions—the way we interact with symbols.

Nuclear weapons are symbols first. So, when we think about nuclear weapons, we actually shouldn't imagine some dark, gigantic shape looming over us. When we think analytically, we shouldn't think of nuclear weapons as the most powerful tools available. Rather, we should conceive of them as a crude wooden statue of sorts—a statue that has a treasured place all its own in a niche of stone where it sits quietly, waiting for the afternoon light to play across its crudely carved features. At night, parents whisper its name in warning to small children to settle them down to sleep. The folk of the village carry it solemnly at the head of processions on holy days. Most nights, candles brought by worshippers burn at its feet. This is how we ought to think about nuclear weapons: as the idol that people put their faith in despite the fact that it is no more than wood hewn into a rough resemblance. Our constant fixation on the weapon with its destructive power blinds us to the way in which nuclear weapons are actually used: as symbols.

THE RISKS OF SYMBOLIC THINKING

You might ask, "So, what's wrong with using nuclear weapons as symbols part of the time? You've said that symbols are an important part of everyday human activity. They add meaning and richness to our lives. They're a part of most things we do. So, why shouldn't they be a part of the thinking we do in order to keep our nation safe?"

It is true: symbols enhance the meaning in our lives. But in discussions of national security, clarity may be more important than enhanced meaning. It's fine to use symbols in love, religion, storytelling, or thinking about the arc of your life. But when you're fighting to stay alive, don't you need to be as realistic as you can possibly be? In a crisis where real risk is involved, a concrete and factual understanding is more likely to avoid danger than intense feelings. Feelings, after all, have been known to cloud perception and make judgment more difficult. This is exactly what happened with battleships in the early twentieth century. Battleships became symbols, and ultimately, that symbolic image got in the way of the realistic estimates of their value that national leaders needed to make.

There are three specific risks involved in weaving symbols into the fabric of your defense thinking: the risk that a symbol will fail to "work" on others, the risk that symbols will change their meaning, and the risk that they will work, in a sense, too well—that they will blur your own perception of reality. Each risk exposes us to peril.

NOT UNIVERSAL

The first risk flows from the fact that symbols are not universal. Some people may believe strongly in a particular symbol, but others may not. You would be hard-pressed to get me to risk my life for the Italian flag, for example. The Italian flag is just some other country's flag to me. But to an Italian, risking his life to protect an Italian flag might seem altogether natural and right. Different people see symbols differently. Some feel the power of one symbol, while some feel the power of another. The symbolic power of an object, after all, isn't something in the object. Its symbolic power resides entirely in our minds (or hearts). Objects don't actually vibrate with a mysterious power. It is only our feelings about them that vibrate in this way. So, while I may be moved by a symbol—and even take action based on those feelings—you may be completely unmoved.

One strong piece of evidence that relying on symbols for defense is risky is that ancient civilizations almost never relied on these kinds of psychological defenses. Symbolic defenses have certainly been possible since the beginning of time. The fact that so few (if any) communities have relied on them is telling. Have you ever heard of a ruler who decreed, "We have no need of walls! Our fearsome god is known far and wide! All fear his terrible wrath and unremitting vengeance! Henceforth, we tear down our walls and rely on Baal Shatra!"? Of course not. Because everyone knows that the meaning and power of a symbol varies from person to person.

The effectiveness of nuclear deterrence—using the symbolic power of the weapons to frighten and dissuade our adversaries—depends on our adversaries feeling the same way about their symbolic value as we do. But people feel differently. This is an undeniable weakness of relying on nuclear retaliation for safety.

CHANGEABLE MINDS

The second problem with using symbols for protection has to do with the fact that human beings have flexible minds. A friend of mine once told me that she never let her sheep make the same mistake three times. "Sheep are so stupid," she said, "that if they make a mistake three times, it gets fixed permanently in their heads. You can never correct them. Once their minds learn a pattern of behavior, they will repeat it for the rest of their lives." I don't know if that's actually true, but it is certainly true that human beings occupy the other end of this spectrum. Unlike my friend's sheep, the human mind is enormously flexible. We can give up old ways of thinking at any time in our lives. Think of the number of people who convert to a new religion or start a new exercise regimen and take up an entirely new way of believing and living.

In U.S. society, flexibility is almost always viewed as a good thing; but in certain circumstances, flexibility is not a benefit. If you are relying on mind-based defenses, for example, you don't want people to be able to change their minds. You want them to always believe in the symbol of power you're using to prevent them from attacking. But human beings can. Just as it's possible for someone who is a lifelong agnostic to suddenly become a born-again Christian, it is also possible for someone who has believed in nuclear deterrence all their life to suddenly be gripped by a conviction that it doesn't matter. Symbols have power only in our minds, and our minds are changeable. This is an inherent weakness when it comes to relying on symbols for security.

Psychological defenses can also fail in another, more radical way. Not only can people change their minds; they can also lose them. There is no mind-based defense that works against someone who is overwhelmed by passion, who is in the grip of a mental compulsion, or has simply lost all touch with reality. The ancient Norse warriors that were called "berserkers," for example, fought in a trancelike fury and would have been immune to mind-based defenses. Relying on psychological defenses means that you are at the mercy of your adversary's mental health.

SELF-DELUSION

The third danger inherent in relying on symbols is that they can warp your own perception of reality. Symbolism makes an object more meaningful, but it also changes the way we see that object, distorting and exaggerating it. Seen through the lens of symbolism, objects may appear bigger than they actually are.

This is what happened to the people who saw a symbol when they looked at battleships. For them, battleships pulsed with meaning. They thought of past victories, they looked at the impressive size, and the feelings they felt made them love battleships. When they looked at battleships, they felt rather than seeing with coldly objective eyes. As a result, the reality of war at sea was hidden from them because their eyes were fixed on an image inside their heads. They ended up making foolish choices that led to disaster.

This power that symbols have to warp our perception can be temporary—a passing problem—but it can also harden in our minds. Look at the world through distorting lenses long enough and eventually the distortion becomes "reality" for you.

Think about money again. No one who stopped and thought about it would claim that money has intrinsic value. On a desert island, a pile of $100 bills would be useless (except, perhaps, as kindling). But by handling money every day, by worrying about it, obsessing over it, yearning for it, we fool ourselves into believing that it is actually something real. People in the grip of this belief dedicate their lives to money, sacrificing their health, their time with their families, and sometimes their happiness for it. People lose their lives over money, dying in dark alleys because they won't surrender it. Somehow, something inside the human brain allows us to convert a symbol into something that seems solid. Familiarity and constant handling slowly take the dreamlike feeling and change it into an apparently concrete fact. But just as mist cannot be spun into gold, symbols are always symbols.

Symbols are powerful forces in the lives of human beings. They shape the meaning at the center of our life stories. But symbols bring risks in their wake. They have uneven effects from one person to the next—even in one individual their power may not persist, and they can distort our own perceptions.

MISTAKING SYMBOLS FOR REALITY

Clearly distinguishing between symbol and reality is essential. When governments confuse weapons with symbols, the consequences can be grave. It can lead, in fact, to fatal mistakes. To see what happens when governments fail to distinguish symbols and reality during a crisis of the highest order, examine the Cuban Missile Crisis. The Cuban Missile Crisis was described by one of the participants as a confrontation "which brought the world to the abyss of nuclear destruction."[166]

One of the critical puzzles for President Kennedy and his aides during the crisis was understanding Premier Khrushchev's motives for putting nuclear missiles into Cuba.[167] A great deal was riding on their assessment. If they misjudged Khrushchev's motives, they might unintentionally present him with demands he could not meet. Faced with unreasonable demands, he might dig in his heels, make his own unreasonable demands in return, and the crisis could spiral out of control.

But Kennedy and his aides, despite the importance of getting their assessment of Khrushchev's motive right, were in doubt. Khrushchev's behavior seemed nonsensical to them. They could not understand the gap between the potential gain for the Soviet Union and the risk the Soviets were running. Viewed from Washington, putting forty-two missiles into Cuba didn't seem like much of a payoff when weighed against the danger of sparking a confrontation that could lead to war. Because the number of missiles was relatively small, several people, including Secretary of Defense Robert McNamara and presidential aide Theodore Sorensen, argued (on the first day of the crisis) that the missiles in Cuba would hardly affect the strategic balance at all. Other aides argued that the missiles would have some effect, but almost no one believed they were militarily decisive. After all, as Kennedy would say at one point during the crisis about the balance of nuclear forces, "What difference does it make? They've got enough to blow us up now anyway."[168]

The Kennedy administration's estimate of Khrushchev's motives would shape U.S. actions during the crisis and perhaps determine whether it could be resolved peacefully or would spiral into nuclear war. But they couldn't seem to grasp what the Soviets were up to. "If they [the missiles] did not matter strategically, why would the Russians put them in? Were they unhappy with their ICBMs [intercontinental ballistic missiles]? Khrushchev was running a major risk. What did he think he could get out of deploying these missiles in Cuba?"[169]

In retrospect, it's possible to see that the reason the Soviet decision seemed inexplicable is because Kennedy and his advisors were looking at the nuclear missiles as weapons, while Khrushchev was seeing them as symbols. Instead of measuring the military benefit that Khrushchev might get from putting forty-two medium- and intermediate-range missiles into Cuba, they should have been trying to understand how—as symbols—nuclear weapons might seem valuable to the Russians.

With the clarity of hindsight and with a new appreciation of how powerful the symbolism of nuclear weapons can be, we can guess (although it is impossible to know) that Khrushchev saw the symbolic value of the missiles as their most important feature when he made the decision to put them into Cuba. At a conference held years after the crisis, Soviet participants "intimated that Khrushchev installed the missiles in Cuba to avenge the sense of shame and humiliation he experienced as a result of the installation of the Jupiters in Turkey.[170] Fydor Berlatsky, Khrushchev's speechwriter, commented, "I'm not sure Khrushchev thought out the aims. From my point of view, it was more an emotional than rational decision."[171] There is a case to be made that Khrushchev felt that the medium-range, nuclear-tipped missiles the United States had deployed in Turkey were a humiliation for Russia. Their special emotional significance for him is documented. Could Khrushchev have felt that the missiles in Turkey symbolized Soviet inferiority and humiliation and that placing missiles in Cuba—even if it had little military impact—would symbolically redress the balance of humiliation?[172]

This notion is supported by the way Khrushchev seemed to emphasize nuclear weapons' value as symbols in other contexts. As soon as the Soviet Union had missiles capable of reaching the United States, Khrushchev felt emboldened to shift to a more aggressive foreign policy. He didn't wait until the military balance of nuclear forces was equal. Once he had a handful of intercontinental missiles, he acted as if the situation had changed. If Khrushchev had viewed nuclear weapons as exclusively military tools, wouldn't he have insisted on a cautious foreign policy until the Soviet Union had achieved military parity with the United States in nuclear weapons? But because he saw them as symbols, the actual numbers of missiles didn't matter.

And Khrushchev's decision to put missiles in Cuba did mimic the symbolism of the missiles in Turkey to a remarkable degree. Placing medium-range nuclear missiles in an Allied country, right on America's doorstep—Cuba is only ninety miles from Florida—neatly mirrored the U.S. medium-range missiles in an Allied country right on Russia's doorstep.

The inability to understand Khrushchev's thinking left Kennedy without a clear idea of where Khrushchev's red lines might be, what his ultimate goals were, and therefore how best to climb back down off this perilous cliff edge they were both on.

There is, in the United States at the heart of our nuclear weapons policy, a confusion. This confusion is seventy years old. There are tangles in logic that arise from this confusion that are now deeply embedded in national security policy. This is a very serious concern. We now see the complications that using objects as both weapons and symbols can cause. We now understand why people said apparently contradictory things in the past. We now see concrete and vivid examples of the danger such confusion causes (in the Cuban Missile Crisis, for example). But it is not clear that even today the government officials in all nine nuclear-armed states keep this dual nature of nuclear weapons firmly in mind.

Intellectual clarity is essential when the stakes are very high. If you fail to understand that nuclear weapons are not one thing but two, that there is not only the machine but also the large, ghostly image of the mushroom cloud rising in your mind's eye, it makes handling them in a sensible way almost impossible.

CONCLUSION

Nuclear weapons are important around the world. They influence people's behavior; they exert a gravitational pull on world events. They stand near the center of important national doctrines and institutions. Yet there is a good deal of evidence that they are not very good weapons. How is this possible? How could they be both things at once: basically, useless weapons and enormously influential objects in world politics?

It is possible because nuclear weapons are two things: the clumsy and embarrassingly unhandy weapons and the shining and awe-inspiring symbols. Nuclear weapons, like chariots, like mounted knights, like battleships, and like many other weapons before them, have become the currency of power. They are internationally recognized symbols of national greatness and military might. But as we have seen from looking at how symbolism works, their status as the currency of power has nothing to do with whether they are actually useful or not. Anything can be a symbol, even a piece of plywood spray painted neon orange. So, it is perfectly possible, through a series of mistakes and misperceptions, for a blundering weapon to become the central currency through which nations handle, evaluate, and use power. The meaning of an object can expand and expand like a giant balloon, swelling ever larger in the imaginary world we see in our mind's eye. But that doesn't mean the object is more and more useful. Meaning is not connected to practical value. Today many leaders and ordinary people believe that nuclear weapons are the gold standard of weapons. But it turns out the coins of that currency are made from lead, not gold.

SYMBOLS

Clarity about nuclear weapons is deadly serious. The consequences of getting caught up in symbolic meaning were serious when battleships were used as symbols: Germany lost a war, and Great Britain lost an empire. But the consequences of being unable to see nuclear weapons as they really are could be much, much worse. Mistakes about nuclear deterrence could lead to nuclear war, and nuclear war could mean the destruction of a great deal of the ancient and mighty fortress that is our civilization.

It is not wrong to use symbols to give meaning to our lives. But it is vital to be able to see past the symbol to the reality when survival is at stake. Reality is harsh and remorseless: it bends for no symbol and forgives no foolish belief.

It is not realism to confuse symbols with reality.

161 Clark, *The Greatest Power on Earth*, p. 210.
162 Shu Guang Zhang, "Between 'Paper' and 'Real Tigers': Mao's View of Nuclear Weapons," in *Cold War Statesmen Confront the Bomb: Nuclear Diplomacy Since 1945*, ed. John Lewis Gaddis, Philip H. Gordon, Ernest R. May, and Jonathan Rosenberg (Oxford: Oxford University Press, 2005, 1999), p. 213. See also Jeffrey Lewis, "China's Nuclear Idiosyncrasies and Their Challenges," *Proliferation Papers*, no. 47 (November-December 2013).
163 Rosendorf, "John Foster Dulles' Nuclear Schizophrenia," p. 63.
164 John Mueller, "Epilogue: Duelling Counterfactuals," in *Cold War Statesmen Confront the Bomb: Nuclear Diplomacy Since 1945*, ed. John Lewis Gaddis, Philip H. Gordon, Ernest R. May, and Jonathan Rosenberg (Oxford: Oxford University Press, 2005, 1999), p. 278.
165 Ludwig Wittgenstein and William James alert.
166 Robert F. Kennedy, *Thirteen Days: A Memoir of the Cuban Missile Crisis* (New York: W. W. Norton & Company, 1969), p. 23.
167 In his meeting with the Joint Chiefs of Staff on Friday, October 19, 1962, for instance, President Kennedy focused on this question first. Ernest R. May and Philip D. Zelikow, eds. *The Kennedy Tapes: Inside the White House During the Cuban Missile Crisis* (New York: W. W. Norton & Company, 2001), p. 111.
168 Ernest May and Philip Zelikow, eds., *The Kennedy Tapes: Inside the White House During the Cuban Missile Crisis—the Concise Edition* (New York: W. W. Norton & Company, 2001), p. 62.
169 Trachtenberg, *History and Strategy*, pp. 245–248.
170 In his memoirs, Khrushchev talked about the Soviet Union being "ringed" with U.S. bases and nuclear weapons that "threatened us" and how the emplacement of missiles in Cuba was "nothing more than giving them a taste of their own medicine." Quoted in Blema S. Steinberg, "Shame and Humiliation in the Cuban Missile Crisis: A Psychoanalytic Perspective," *Political Psychology*, 12, no .4 (December 1991), p. 667.
171 Steinberg, "Shame and Humiliation in the Cuban Missile Crisis," pp. 665–668.
172 "A successful deployment of missiles in Cuba, moreover, would have been a symbolic demonstration of Soviet parity in the political, if not in the military, sense." Steinberg, "Shame and Humiliation in the Cuban Missile Crisis," pp. 665–666. The importance of humiliation in the deterrence failure of the Cuban Missile Crisis is ratified by none other than President Kennedy himself. Notably, the lesson he drew from the Cuban Missile Crisis was that you should never force your adversary to choose between nuclear war and humiliation. (Whether Kennedy was talking about Khrushchev's humiliation at being forced to withdraw Soviet missiles from Cuba or his own realization that the sudden discovery of operational missiles in Cuba would have been a stunning humiliation for the United States and for him and would probably have led to his losing the election of 1964 can't be known.)

10. DETERRENCE THEORY

It is possible because nuclear weapons are too dangerous to keep; deterrence theory is doubtful and untestable.

Some nuclear weapons advocates admit that nuclear weapons are not very good weapons. And they admit that nuclear weapons are mostly symbols. But they argue that none of that actually matters. They might say, "The fact is that no one really wants to use nuclear weapons. Nuclear weapons' only real purpose is to deter. That is why we need these weapons, why we have to keep them. Without nuclear weapons, we couldn't deter the Russians or the Chinese. Deterrence is the essential thing. It buttresses our alliances, and those alliances maintain the world order."

Some nuclear weapons advocates even admit that nuclear deterrence is built on very little factual evidence. But they argue that it doesn't matter if you arrive at a theory by intuition or careful, step-by-step experimentation. "The proof," they might say, "is not in the way you build a theory. The proof is in the way it works. The fact is, there hasn't been a major war in seventy-five years. Nuclear deterrence has kept the peace in a truly remarkable way. And there is a case to be made that as long as we keep our nerve and maintain a certain level of nuclear weapons, deterrence can work for as long as we need it to." And they argue this point with surprising conviction. They seem quite confident that nuclear deterrence does work—and works robustly.

So, the question this chapter tries to answer is this: is that confidence justified? Is the case for nuclear deterrence so strong and so sound that it can make us believe that nuclear deterrence is safe, even though real evidence is almost entirely lacking? Sometimes a theory is so obviously true that we rely on it despite not having very much proof. Is it possible that nuclear deterrence theory is one of these theories?

*   *   *

Some nuclear weapons advocates claim that nuclear deterrence has never failed. There has been no nuclear war for more than seventy years, they point out, so questions about the reliability of nuclear deterrence are unfounded. From one perspective, the case that threatening punishment works to deter seems self-evident. One early exponent of deterrence wrote, "That the fear of punishment can deter is shown … vividly by its efficacy in the training of animals."[173] And that's true. You can, for example, train a dog not to get up on the sofa by hitting it with a slipper. Hit the dog a few times and pretty soon just the sight of the slipper will likely make him slink away to his corner. Nuclear weapons advocates then say, "If you can deter a dog with a slipper, surely you can deter a human being with nuclear weapons. Human beings are far smarter than dogs. And nuclear weapons are far more frightening than a slipper." It seems like a telling point. If it is that easy to make deterrence work, isn't it likely to work well in most circumstances?

Nuclear weapons advocates are so sure. The fact is, it's a little unnerving how firmly they argue and how sure they seem. Could they possibly be wrong when they seem so sure? Let's take a closer look at the theory of deterrence and see if their confidence is justified.

PERFECTION

Before we begin, it's worth noting that the bar for nuclear deterrence is quite high. Normally if a theory works in the great majority of cases, or even just in most cases, it counts as a useful explanation of the world. A theory about what causes brain disease that is right eighty-six percent of the time would probably count as a useful theory. But because the consequences of nuclear war are so severe, the demands on nuclear deterrence theory are much greater. Nuclear deterrence simply cannot fail—the consequences are too great for us to allow it to happen. Even one failure could be catastrophic. So nuclear deterrence has to be perfect; it has to work every time. This is a demanding requirement. And it turns out that when you compare nuclear deterrence to other kinds of deterrence, the comparison seems to suggest that nuclear deterrence is unlikely to get over the bar.

Remember that nuclear deterrence is only one kind of deterrence. There are many different kinds. Deterrence is a type of threat, and threats can be used in a variety of circumstances—for example, threatening to spank a child, threatening to take someone's driver's license away, threats used in ancient times to starve a city unless it surrendered, threatening to punish certain crimes with the death penalty, and so on. Each of the many different types of deterrence has different characteristics and a different likelihood of succeeding. But they all share one characteristic: none has a perfect record of success. Sometimes children do misbehave. Sometimes cities elect to starve. Sometimes people drive when they are drunk. Of all these different kinds of deterrence, none works perfectly. Even when the evidence shows that a particular kind of deterrence seems to work much of the time, there are still occasions when it fails. So, if nuclear deterrence is like these other forms of deterrence, if what applies for them also applies to threats with nuclear weapons, then it seems probable that nuclear deterrence will also fail some of the time.

Nuclear weapons advocates sometimes argue that nuclear deterrence will work better than these other types of deterrence because the penalty that you pay if nuclear deterrence fails is so much greater than the penalty with these other kinds of deterrence. And they have a point. The destruction of civilization and the loss of, say, two hundred million lives could be called the ultimate penalty. But this argument actually highlights a weakness, because for a murderer, the death penalty is also the ultimate penalty. There are real parallels between the severity of nuclear war for civilization and the severity of the death penalty for an individual. What greater penalty could there be for an individual than death? Does a murderer care if life goes on after his death or not? His life has ended. For him, the world has come to an end. The fact that the death penalty regularly fails to deter, therefore, is particularly troubling. The death penalty is not a perfect deterrent. It turns out that people regularly fail to be deterred by the threat of paying the ultimate penalty.

This tendency to fail in other kinds of deterrence casts a shadow over nuclear deterrence. If other kinds of deterrence cannot work every time, how can we believe nuclear deterrence will work every time? If deterrence using other means is imperfect, why should we expect deterrence with nuclear weapons to be different? No, no, no, no, say nuclear weapons advocates. Nuclear deterrence is different. It has specific characteristics that make it more likely to succeed. They put forward three arguments for why nuclear deterrence is different from other forms of deterrence: because it carries with it a unique power to hurt, because nuclear war is so clearly irrational that no one in their right mind would choose it, and because of the overpowering fear of nuclear war. These three arguments are distinct—they do not interlock or support one another. Different groups of nuclear weapons advocates rely on each of them. So, let us examine each in turn.

THE POWER TO HURT

Thomas Schelling, a Nobel Prize-winning economist who is sometimes called the founder of nuclear deterrence theory, argued that the threat of enormous harm that nuclear weapons create is a powerful tool for influencing others. Schelling argued that nuclear weapons' unprecedented ability to devastate and punish could be used to ensure that deterrence works reliably. In his *Arms and Influence,* Schelling writes eloquently and at length about how the "power to hurt" makes deterrence effective. "It is the power to hurt, not military strength in the traditional sense, that inheres in our most impressive military capabilities at the present time."[174]

The difficulty with the power to hurt that Schelling talks about is that many of the leaders who make decisions about war and peace are like Ronald Niedermann. Ronald Niedermann is a character in the very popular trilogy by Stieg Larsson that begins with *The Girl with the Dragon Tattoo.* Niedermann is the half-brother of the heroine, Lisbeth Salander. He is abnormally large and strong, and he is a member of a gang that sells drugs—in other words, he is a very bad guy. He kidnaps Salander's friend, and in a key scene of the second book, Niedermann faces off against a professional boxer, another friend of Salander's, who has come to rescue the woman who has been kidnapped. The boxer, while he's impressed with Niedermann's size, is still confident. He's competed at a very high level in his weight class and is sure he has the ability to fight any man. However, during the fight, a peculiar thing happens. Despite the boxer landing several wicked punches—blows that would have knocked out a lesser man—Niedermann seems unfazed. He shakes his head, grunts, and then continues fighting. It's almost as if he can't feel he is being hit.

And it turns out that is exactly the problem. Niedermann has a rare neurological condition called "congenital analgesia," a condition in which a person cannot feel—and has never felt—pain. In the fight with Salander's boxer friend, Niedermann doesn't crumple under the blows because he literally cannot feel them. In any fight, a person with congenital analgesia is less likely to give in because he will not feel any pain from the blows that he receives.

Schelling's contention that the power to hurt is power over-looks the fact that many national leaders are like Niedermann: they have congenital analgesia when it comes to civilian casualties in war. This is not to say that they did not start out life with the normal amount of human sympathy for the suffering of others. But in wartime, they feel a responsibility to the needs of the state that outweighs this sympathy. When war is fought and the necessity to preserve the state is challenged, leaders are not likely to put a high value on civilian suffering. As the examples already cited demonstrate, wartime leaders are capable of enormous callousness when it comes to civilian lives. Chiang Kaishek drowning his own civilians, Joseph Stalin letting ordinary Russians starve in Leningrad, Winston Churchill allowing British cities to be bombed—the list of war-time leaders who found ways to live with the deaths of civilians is quite long.

It seems probable that most leaders in wartime are connected to citizens similarly to the way that Niedermann feels pain. They have a disinterested, abstract understanding of what is going on; they understand that the body they are a part of is being injured, but they do not actually feel the pain in any visceral, immediate way—at least, any pain they feel doesn't overwhelm their assessment of the strategic balance in the conflict.

Despite the popularity of Schelling's work with people who call themselves realists, it's not clear that Schelling's description of the "power to hurt" is realistic. As the nuclear scholar Robert Jervis pointed out, "One could not have coerced Pol Pot by threatening to destroy his cities."[175] Nuclear weapons mostly hurt civilians, but civilian deaths rarely, if ever, determine whether a leader continues a war.

RATIONALITY

So, let us turn to the second way that nuclear weapons advocates claim that nuclear deterrence cannot fail: the assertion that no rational person could think about the consequences of a nuclear war and then choose to fight one.

For some reason, the assumption of rationality is deeply lodged in nuclear weapons thinking. From the earliest days of nuclear strategy, when people first began to think about how best to fight a nuclear war, there seems to have been a powerful inclination to assume that rationality would play a large role in decision-making. From the outset, when game theory and logic models dominated, to the more recent history of rational choice thinking, the lure of imagining that decisions would be made based on reason and cost-benefit analysis has been strong.[176]

This is surprising because there is simple and undeniable proof that human beings are not rational—proof that is widely available and easily accessible. In fact, this proof is so commonly available that it is virtually everywhere that human beings now live.[177] In fact, one piece of this omnipresent evidence is probably in the room with you right now. Look around. See it? If you don't, reach down and gently pat your stomach just above your belt line. Now, using your thumb and forefinger, grasp a little bit of the flesh above your hip on one side or the other of your torso. Did you just forget about nuclear weapons for a moment and think to yourself,

"You know, I should probably lose five pounds"? If you did, you're probably like the great majority of people in the developed world. And that is entirely adequate proof that we are not completely rational.

Consider this: is it rational to be overweight? Is there not a great deal of data showing that being overweight has negative consequences for health? Isn't it the case that the people who seem to live longest are often quite trim and live on relatively low-calorie diets? Now think about some of the habits that contribute to your being a little overweight: smoking, too many sodas, sweet snacks, sitting lazily on the sofa watching videos instead of running and playing outside. I'm not lecturing you. I could really stand to lose about ten pounds myself (maybe fifteen, if I'm honest). What I'm asking is this: if human beings were really rational beings, and if our decisions were controlled by rational choices, would anyone be overweight? There's a great deal of evidence that maintaining an ideal weight is really healthy for you. And that the lifestyle of activity and exercise that's required for achieving such a weight has enormous benefits, too. So, why are so many people overweight? Is it rational? Is it a conscious choice to do things that are not the best for your health? Obviously not.

It is clear that our choices are sometimes controlled by what we think and what we consciously tell ourselves. It is possible for us to decide on a course of action rationally. But much more often, our choices—in eating, most obviously, or in love or in whether to drink alcohol, in fact, in many categories of behavior—are controlled by urges, instincts, desires, and emotions that are very difficult for our conscious mind to control. If a great deal of our behavior is in thrall to subconscious emotions and urges, is it such a stretch to imagine that our choices in wars might be propelled by emotions, too?

What is striking about the argument for rational choice and game theory is that it continues to appeal to nuclear weapons advocates, despite the obvious and commonsense evidence that rationality only sometimes rules our choices. This insistence on a theory of rationality has real-world consequences. One example of the way these assumptions of rationality played out in actual U.S. policy occurred during the administration of President Richard Nixon. In the 1960s, think-tank intellectuals had explored an approach to foreign policy bargaining they called "uncertainty strategy." Schelling and others began to argue that it could be useful to seem to be a little crazy in a nuclear crisis. If you make a threat, according to the theory, and you seem to be a little crazy, the chances that the threat will work increase.

"Uncertainty strategy" eventually made its way into Nixon's thinking, and he decided to use it to try to force North Vietnam to negotiate an end to the war in Vietnam. H. R. Haldeman, one of the president's top aides, picks up the story in his memoir of Nixon's presidency:

> Nixon not only wanted to end the Vietnam War, he was absolutely convinced he would end it in his first year.... "I'm the one man in this country who can do it, Bob."....
>
> He saw a parallel in the action President Eisenhower had taken to end [the Korean] war.... He [Eisenhower] secretly got word to the Chinese that he would drop nuclear bombs on North Korea unless a truce was signed immediately. In a few weeks, the Chinese called for a truce and the Korean War ended.
>
> In the 1950's Eisenhower's military background had convinced the Chinese that he was sincere in his threat. Nixon didn't have that background, but he believed his hardline anti-Communist rhetoric of twenty years would serve to convince the North Vietnamese equally as well that he really meant to do what he said....
>
> The threat was the key.... "I call it the Madman Theory, Bob. I want the North Vietnamese to believe I've reached the point where I might do anything to stop the war. We'll just slip the word to them that, "for God's sake, you know Nixon is obsessed about Communism. We can't restrain him when he's angry ... and he has his hand on the nuclear button"—and Ho Chi Minh himself will be in Paris in two days begging for peace.[178]

The uncertainty strategy assumes that if you are crazy, it will change the cost-benefit calculation of your adversary. Your adversary will calculate that if you are crazy, there is a greater likelihood that you will actually do what you say and, therefore, the safest course is to capitulate. By acting as if you are irrational, so the theory goes, you create a significant advantage.

The problem with this theory is that it assumes that your acts of irrationality exist in a world of rationality. It assumes that while you are acting insane, your adversary is being reasonable. But human beings only act rationally some of the time. The assumption that when Nixon acted irrationally, the North Vietnamese would respond by thinking things through and making a cost-benefit assessment might have been true. If your adversary started acting crazy, you might think harder about what your most rational response would be.

But it could also be that instead of being more reasonable when you act crazy, an adversary might respond by being emotional. Emotions tend to beget emotions. Crazy behavior on your part might well lead to crazy behavior on your adversary's part. Rather than making your adversary more cautious, crazy behavior on your part might lead to an escalating spiral of emotion—anxiety leading to fear, fear leading to threats, threats leading to greater fear, and, in the end, one leader orders a preemptive nuclear strike.[179]

In the event, the strategy did not work. Despite issuing a string of threats in 1969, Nixon was not able to force the North Vietnamese into significant concessions at the peace talks. Despite Nixon's certainty that he could resolve the conflict, the war in Vietnam ground on for another six years. Thousands of U.S. soldiers, North Vietnamese soldiers, and civilians died during those years—a high price to pay for being wrong. But the costs could have been much higher. If Nixon had tried to rely on the uncertainty principle in a nuclear crisis with another nuclear-armed state—with the Soviet Union, for example—the costs of being wrong could have been catastrophic.

The idea that the risk of nuclear war will call forth untapped reserves of rationality in leaders is a pleasant one. To borrow Ernest Hemingway's phrase, it is, "pretty to think so." But it seems unlikely to be true all the time or perhaps even most of the time. The nuclear weapons advocates who are relying on rationality to protect us from nuclear war are not being realistic.

As Freeman Dyson put it:

> Assured destruction would be a stable strategy in a world of computers. In a world of human beings, it fails to bring stability because it lacks the essential ingredients which human beings require: respect for human feelings, tolerance for human inconsistency, adaptability to the unpredictable twists and turns of human history.[180]

A realistic approach to nuclear war and nuclear crisis, it seems to me, would assume that decision makers will be driven by instinct, buffeted by emotion, and would be susceptible to unreason. It would expect that a crisis involving nuclear weapons would be like other human crises where the stakes are high and the danger is very great—in other words, the pressures of such a crisis would make people more vulnerable to mistakes, emotional outbursts, and irrational behavior.

Insisting on rationality that does not exist is not realism. It is wishful thinking, and dangerous wishful thinking at that. If the assumption of rationality in a crisis turns out to be wrong, millions of innocent people could suddenly and irrevocably pay for it with their lives.

THE POWER OF FEAR

Finally, another explanation advanced by nuclear weapons advocates for why nuclear deterrence might be different from other forms of deterrence (and therefore less likely to fail) is the "power of fear" explanation. Nuclear deterrence cannot fail, they argue, because the image of nuclear war is so frightful, the danger so clear-cut, and the power of fear so imperative that it would be impossible for any leader to ignore the danger. Rationality might not restrain a leader, and the possibility of pain inflicted might not work but fear most certainly would, according to this argument.

This is, in some ways, a more plausible argument than either the pain or rationality arguments. It is certainly true that anyone who has seriously contemplated nuclear war has felt deeply afraid of it. And fear is a much more powerful and fundamental motivator of action than rationality. But even this stronger argument is not persuasive.

Is fear the most powerful emotion? Despite what nuclear weapons advocates might say, fear does not always dominate. Brave men—through training and loyalty to their small group—overcome fear all the time in war. During Roman times, the berserkers of some Germanic tribes were said to feel no fear when they were in their frenzy. Zulu warriors simply took hallucinogenic drugs that suppressed their fear before battle. Belief in an afterlife can overcome fear. A desire for revenge can overcome fear. Sometimes lust for killing can overcome fear. False optimism can overcome fear. Delusional people are not restrained by fear. Drunkards can forget to be afraid. Enraged people can ignore fear. Sometimes starvation and hunger overcome fear. Sometimes a profound sense of humiliation overcomes fear.[181] People motivated by intense love can act bravely for the sake of their beloved. And so on. There are a multitude of emotions that can overcome fear—some for only a short time, others for much longer. But the point is, fear is not the ruling emotion, master of all the others.

The problem with the notion that fear will always restrain aggression when there is a risk of nuclear war on hand is that fear is not king. It cannot restrain all the other emotions all the time. Sometimes fear will be strongest, but sometimes other emotions or instinctual urges will be stronger. We know this is true from the experience of soldiers. We know it from history. We know it from psychology. Fear is not strong enough to command all the other emotions.

Nuclear weapons advocates might respond, "But nuclear war is different. No other fear includes the danger of vast destruction, perhaps even the end of civilization itself." And it's true: nuclear war would include some elements—like widespread radiation poisoning—that human beings have never experienced before. But there is no evidence that this difference invalidates what we know about human nature. In fact, there is clear and incontrovertible evidence that a world leader could look the prospect of nuclear war squarely in the eye and still not be afraid.

Compelling evidence in this regard comes from the Cuban Missile Crisis. Fidel Castro, as leader of Cuba, was concerned as the crisis mounted. The likelihood of worldwide nuclear war was obvious to all observers, and there was overpowering dread and anxiety in the United States, in Europe, in Russia, and in other parts of the world as people waited fearfully to see what would happen. But that is not what was worrying Castro. Castro was worried that Soviet Premier Khrushchev might shrink back from nuclear war. In the early morning hours of October 27, 1962, the twelfth day of the crisis, Castro was awake, drafting a cable for Khrushchev:

> Dear Comrade Khrushchev, Analyzing the situation and the information that is in our possession, I consider that an aggression in the next 24–72 hours is almost inevitable…
>
> If the aggression takes the form of the second variant and the imperialists attack Cuba with the purpose of occupying it, the danger facing all of mankind … would be so great that the Soviet Union must in no circumstances permit the creation of conditions that would allow the imperialists to carry out a first atomic strike against the USSR.

Castro, as befits a man who had been trained by Jesuits, often used intricate reasoning and complex wording. As he drafted this letter, talking with associates, some of them weren't entirely clear about what he was driving at. Finally, one of his aides "blurted out" the question: "Do you want to say that we should deliver a nuclear first strike against the enemy?" Castro replied that that would be too blunt:

> No, I don't want to say it directly. But under certain conditions, without waiting to experience the treachery of the imperialists and their first strike, we should be ahead of them and erase them from the face of the earth, in the event of their aggression against Cuba.[182]

Perhaps the most striking thing about Castro's letter urging nuclear war is that he must have known that in a nuclear war, Cuba would be laid waste. Even if the Russians launched a first strike, and even if that first strike was devastatingly effective, some U.S. forces—both nuclear and conventional— would have survived. In the ensuing war, Cuba would have been a principal target. At the very least, the nuclear missile sites being built in Cuba would be attacked, and probably Cuba's most important cities as well. Urging nuclear war meant urging the destruction of much of Cuba.

Castro looked the vast horror of nuclear war in the face and recommended to Khrushchev, in a roundabout way, that he launch a preemptive first strike with Russia's nuclear forces if war came. He was aware of the stakes and understood that even if the Soviet Union launched a first strike, some U.S. nuclear forces would survive. And yet he urged his ally to launch such a war. How can Castro's letter be explained? If the fear of nuclear war is so great that no leader could face it and not be cowed, how can Castro's actions be possible?

The answer is that his actions can't be explained if fear of nuclear war was controlling his actions. Castro's letter appears to be a direct confirmation that sometimes in a crisis, despite the fear of a nuclear holocaust, a national leader could recommend launching a nuclear war. It is proof that fear does not always rule.

So, even though fear is a powerful emotion, and even though the costs of nuclear war are plain, the notion that leaders will always draw back from the brink because of fear is contradicted both by what we know about human nature and by this extraordinary episode from history. Insisting on the power of fear to restrain nuclear war, when you know that it is not the most powerful emotion, is not realism.

FALSE CERTAINTY

There is an old joke that a Keynesian economist is like a blind man in a dark room trying to find a black cat; a Marxist economist is like a blind man in a dark room trying to find a black cat that isn't there; and a supply-side economist is like a blind man trying to find a black cat in a dark room that isn't there who shouts, "I found it!" False certainty—claiming to have found something you cannot possibly have found—makes the joke funny. But false certainty in foreign relations is both dangerous and foolish.

In 1940, the French high command claimed that no modern tank attack could be launched through the heavily wooded Ardennes region. They were certain of it. France lost the war because of their false certainty. In 1776, the leaders of Great Britain were certain they didn't have to compromise with their American colonies. A collection of distant colonies would be no match for the greatest military power on earth, they assured themselves. Their false certainty resulted in the loss of one of the largest and richest parts of their empire. One of the most profound dangers when thinking about nuclear deterrence is adopting an attitude of false certainty.

IT IS POSSIBLE

Because there is no way to get inside someone's head and shine a light on the workings of their decision-making during a nuclear crisis, because there is no way to measure the determination, credibility, estimates of arsenal size, and so on inside someone's head, it is not possible to know anything definite about how nuclear deterrence works—or doesn't work. Nuclear deterrence is not a science; it is an art.

I once sat with a very senior government official who had dealt with nuclear weapons at the highest level (including getting one of those phone calls in the middle of the night that a nuclear attack might be underway). He is a wise and considerate person and one that I respect a great deal. However, I was surprised to hear him say, "Well, we know that deterrence doesn't work [in situation x], but it will work [in situation y]."[183] The authority of his past experience and his former position at the highest levels of government gave his voice a weight and certainty that made me, for a moment, wonder if he knew exactly when nuclear deterrence would and wouldn't work. But realism also consists in admitting what you don't know. Despite my admiration for him, I am sure he was wrong. It is not possible to know when a nuclear deterrence threat will work and when it will not.

With nuclear deterrence, it is not possible to be certain. In any given situation, you cannot know whether nuclear deterrence will work or fail. The map of deterrence attempts is not divided into areas that are clearly and sharply marked—some labeled "will work," others labeled "won't work." Nuclear deterrence is all the empty place on the map where medieval scholars used to write, "Here be dragons." The entire map is shrouded in darkness.

Based on what we know of human nature, it seems plausible that nuclear deterrence will work—some of the time. And all the different situations where advocates claim it will work can indeed be successful use of deterrence—some of the time. Nuclear deterrence can prevent wars—some of the time. It can give you diplomatic leverage—some of the time. It will secure your treaties—some of the time. It is the ultimate guarantee of safety—some of the time.

But the claim that nuclear deterrence can work every time cannot be true, for two reasons. First, we just don't know that much about how nuclear deterrence works. Nuclear deterrence happens inside the black box of the mind, and we have no tools that can get in there and measure accurately.

For example, imagine that leader A has said that if leader B annexes territory in country C, then leader A will launch a nuclear retaliation in response. Deterrence theory says that fear of nuclear war is a significant factor influencing the effectiveness of a deterrence threat. How do you measure the level of fear of nuclear war in leader B's mind? What instruments do you use? What numbers can you write down? Nuclear deterrence theory says that the credibility of the person threatening to retaliate with nuclear weapons has an impact on whether the deterrence threat works. How do you measure B's assessment of A's credibility? Is there a "credibility assessing" portion of the brain that has been identified? Can we use scans or probes to measure the activity in that part of the brain? And if we could measure "activity," how does one correlate activity in that part of the brain with a leader's credibility? Is there a machine we can use where a reading of six clearly indicates a low level of belief in an adversary's credibility and a reading of twenty-eight means a high level of belief?

Some nuclear deterrence theorists work from past events, trying to create datasets of successes and failures when deterrence seems to have worked. But, although it may be useful, this is an exercise in guessing and estimating. Action A occurs, and then a leader takes action B. Are the two connected? Did A cause B? We have no way of knowing with any certainty. Various people may feel certain that they do, but that is not objective proof. That is no more than a hunch or an intuition. There is no calculation that can be made to measure the causative effect of A on B. No scientific measurement has gone on. Are we really comfortable risking the lives of billions of people on hunches?

So, the first reason that we cannot be certain about nuclear deterrence is that it is a field with very few real facts. The second reason is, if anything, more compelling. Nuclear deterrence cannot be trusted because human beings are unpredictable. We are the variable in the equation. Our moods and emotions change. Optimistic and feeling unbeatable one day, we are downcast and overwhelmed the next. This means that the best we can say, in any given situation, is, "Nuclear deterrence has a better-than-average chance of working in this situation. But there are no guarantees."

Imagine a crisis between a large nuclear-armed state and a smaller one. The small state has a fledgling nuclear arsenal. It has never, in fact, successfully test-fired one of the long-range missiles that are intended to deliver its nuclear warheads. The larger state has weapons that have been thoroughly and successfully tested for decades. The smaller state delivers a nuclear threat. Most nuclear deterrence theorists would argue that this threat ought to fail. The disparity in the sizes of the nuclear arsenals and the difference in the reliability of the means of delivery ought to make the smaller state's threat certain to fail.

But the fact is, the small state's threat could succeed. Leaders lose their nerve all the time for a multitude of reasons. We have already seen several cases of nuclear threats that ought to have worked—like the existence of an

Israeli nuclear arsenal and its implicit threat during the 1973 Middle East War—that failed. In any crisis, nuclear deterrence can fail—even in the most unlikely of circumstances. And in any crisis, nuclear deterrence can also succeed—even when experts predict that it shouldn't. With nuclear deterrence, there are never any guarantees.

In poker, there is no way to scientifically predict whether a bluff will work. There may be factors that often point to a bluff working, but there is no way to use calculations to know with certainty when it will infallibly work. Nuclear deterrence is not chess. It is not logic. It is not calculation. It is gambling. We can only know as much about nuclear deterrence as we know about betting in poker.

So, we have to treat nuclear deterrence carefully—like an explosive chemical compound that is unstable. Most times the compound will probably work. But we can't rely on it always working. Human beings have emotions, and no matter how strongly we set our conscious minds to control our emotions, sometimes our emotions overwhelm our better judgment and spark actions we didn't plan. Nuclear deterrence is not controlled or certain. It is always a gamble.

CONCLUSION

So, we have examined the theory of nuclear deterrence and the confident way that nuclear weapons advocates talk about it. And they do often speak as if they have found the black cat. But they haven't. There is no way to get a measuring device inside the human brain to measure the many different factors that go into a nuclear deterrence decision, and no leader has ever had his or her brain measured in this way during a crisis. There is no reliable evidence about what makes nuclear deterrence work and what makes it fail.

Nuclear deterrence theory is flawed from end to end. Leaders do not feel the deaths of civilians sufficiently to make deterrence failure-proof. The contention that rationality will hold human beings back in times of crisis is contradicted by human nature. The hope that fear will prevent nuclear war is based on the false assumption that fear is always the strongest emotion. And every time you employ nuclear deterrence, even in the most tried and true circumstances, there is a real chance that it will fail.

173 Franklin E. Zimring and Gordon J. Hawkins, *Deterrence: The Logical Threat in Crime Control* (Chicago: The University of Chicago Press, 1973), pp. 1–2.
174 Thomas Schelling, *Arms and Influence* (New Haven, CT: Yale University Press, 1966), p. 7.
175 Robert Jervis, "Deterrence and Perception," in *Strategy and Nuclear Deterrence*, ed. Steven E. Miller (Princeton, NJ: Princeton University Press, 1984), p. 59.
176 For more on rational choice, see Ithak Gilboa, *Rational Choice* (Boston: The MIT Press, 2012); Michael Allingham, *Choice Theory: A Very Short Introduction* (London: Oxford University Press, 2002); and Jeffrey Fiedman, ed., *The Rational Choice Controversy: Economic Models of Politics Reconsidered* (New Haven, CT: Yale University Press, 1996).
177 Except for a few situations in which human beings do not have adequate control over their circumstances, like South Sudan, for example.
178 H. R. Haldeman, *The Ends of Power* (New York: Times Books, 1978), pp. 82–83. It is worth noting that Nixon's belief that nuclear threats were key to ending the Korean War are challenged by most historians and now considered doubtful.
179 Hannah Arendt's comment on violent action applies as well, it seems to me, to violent threats. "Action is irreversible, and a return to the status quo in case of defeat is always unlikely. The practice of violence, like all action, changes the world, but the most probable change is to a more violent world." Hannah Arendt, *On Violence* (New York: Harcourt, Brace & World, 1969), p. 80.
180 Freeman Dyson, *Weapons and Hope* (New York: Harper & Row Publishers, 1984), p. 245.
181 A sense of humiliation is an underappreciated cause of deterrence failure. There is good evidence for it playing a key role in the deterrence failures of the Middle East War of 1973, the Falkland Islands War, and the Cuban Missile Crisis. See works by Richard Ned Lebow and Janice Gross Stein, and especially Blema S. Steinberg, "Shame and Humiliation in the Cuban Missile Crisis," pp. 653–690.
182 Dobbs, *One Minute to Midnight*, pp. 203–204.
183 This is a quote from a conversation with a widely respected, universally liked former government official who is considered one of the leading experts on nuclear weapons in the world today. It is a view I have also heard in conversation with other nuclear deterrence experts many times.

11. DETERRENCE REALITY

*It is possible because the facts show that
nuclear weapons are too dangerous to keep.*

We have examined the theory of nuclear deterrence, but as realists, what we're really interested in is the facts. So, let us delve into the historical record. One of the clearest effects of holding and believing the ideas that make up the nuclear mindset is a tendency to deny the facts. Nuclear deterrence has obviously failed in the past, but nuclear believers (some, not all) contend that the record of deterrence working is unblemished. To do this, they have to hide from some rather uncomfortable evidence. If nuclear believers were really realists, they would not behave this way. Realists welcome the fact-stream of daily events that make up history. They dip into it and examine the individual drops with interest. They like the details. Idealists are the ones who disdain the everyday and focus instead on the "big picture." "Details are for accountants," the idealists say haughtily; they see the grand sweep of history. Given that realists like facts rather than grand visions, you might imagine that nuclear "realists" love a good factual debate about nuclear weapons history. But you would be wrong. They show a curious unwillingness to get involved in the details of the evidence for nuclear deterrence. Their greatest hero is Thomas Schelling, a man who disdained evidence and whose work is almost entirely untethered from actual events.

The record of nuclear deterrence in Cold War crises is an important proving ground for whether nuclear deterrence can be safe and reliable. But nuclear believers consistently misinterpret or ignore that history. This is disquieting because the facts from Cold War crises reveal worrisome discrepancies.

IT IS POSSIBLE

To give a sense of the problem, let's look at just a sample of the cases where problems exist. More cases could be brought forward and argued, and each of the cases here could be discussed at greater length. But these short thumbnail sketches are enough to highlight the scope of the problem.

BERLIN CRISIS OF 1948

Serious issues were at stake in the confrontation over Berlin in 1948. The United States, the United Kingdom, and France on the one hand and the Soviet Union on the other had very different ideas about how to treat Germany now that World War II was over. Negotiations over the matter got nowhere. Finally, the Soviet leader, Joseph Stalin, decided that he would try to get his three former allies to see things his way by cutting off road and rail access to West Berlin.

According to deterrence theory, Stalin's decision is peculiar. In 1948, the United States had a monopoly on nuclear weapons. If the threat of nuclear war has the power to influence behavior, one would think that Stalin would not have risked a confrontation with the United States when the balance of nuclear forces (some to zero) was so much in favor of the United States. And in any crisis, war is always a possibility. Cutting off access to Berlin risked war.

Stalin may have felt that he was in charge, that a war would not happen as long as he didn't order it. But confrontations at such close quarters are always risky and unpredictable. There is a famous picture from another, subsequent crisis in Berlin that occurred in 1961 that illustrates the risk. It shows U.S. and Soviet tanks facing off in the street near the divide between the Soviet zone and the U.S. zone. The tanks, barrels trained on each other, stand no more than one hundred yards apart. The picture is a vivid reminder that when confrontations take place, when emotions are high, a single mistake by a tank commander, by a trigger-happy soldier, or even a shot from a civilian trying to provoke a clash could easily result in a fire fight, casualties, the spread of fighting, and eventually war. Allowing men with live ammunition to confront their opponents (also with live ammunition) was a high-risk strategy.[184] You would think that the danger of nuclear war would have dissuaded Stalin from running this sort of risk. Yet Stalin was not deterred by the United States' nuclear weapons. Deterrence, in other words, failed.

Several months into the crisis, the U.S. Air Force moved B-29 bombers to airfields in Great Britain. (B-29s were the same bombers that had dropped the atomic bombs on Hiroshima and Nagasaki.) The news of the move was "leaked" to newspapers and widely reported around the world. Although the United States was not contemplating the use of nuclear weapons, putting apparently nuclear-capable bombers in the United Kingdom was widely viewed as a clear but unspoken threat of nuclear war.[185]

Two interesting things happened as a result of moving the bombers. First, the Soviets did not rush to resolve the crisis. The confrontation continued for nearly a year before the Soviets eventually relented, and even then, what convinced them the blockade was useless seemed to be the fact that the United States had demonstrated that it could supply Berlin through the air rather than the threat of war.

Second, by the time a year had passed, it was accepted lore inside the U.S. government that the nuclear threat had "worked."[186] Even today, there are nuclear believers who seriously argue that the point of the threat was not to end the blockade but to stop the Soviet Union from escalating the crisis, and that the threat "seemed" to work.[187] But it's hard to think how the Soviets could have escalated the crisis short of starting a war. After all, once you've blockaded the highways and railways, there is little else you can do short of shooting down the planes flying in supplies.

When a threat comes eleven months before the crisis is resolved, it's hard to argue that the threat worked. But nuclear believers don't seem to see the Berlin Airlift crisis clearly. Like old men remembering their high school football glory, the losses are somehow converted in their minds into victories.

KOREAN WAR OF 1950

Because people in the U.S. government eventually came to believe that the threat of moving B-29 bombers to Great Britain had "worked" during the Berlin Blockade, when war came on the Korean peninsula in 1950, it was natural that they would try to repeat their earlier "successful" threat. During the early part of the war, when U.S. and other Allied forces had intervened in the conflict on the side of South Korea and were pushing North Korean forces farther and farther up the peninsula, there was a good deal of concern that either the Soviet Union or the Chinese would join the war to stem the losses on the North's side. Someone apparently suggested that a nuclear threat might be useful, and a threat just like the one in the Berlin Crisis was put into effect.

Again, no explicit words were spoken; no direct threat was issued. But like last time, U.S. bombers were prominently shifted—this time to the island of Guam, where the United States maintains a large military base and from which B-29s could reach targets in Asia. Word of this transfer of military assets again "leaked" to the American press and was again widely reported. The United States was putting nuclear-capable bombers within range of targets in North Korea, Russia, and China.

Subsequently, nuclear believers claimed that the nuclear threat worked. It kept Russia from joining the war on the side of the North Koreans, they claimed. But since the Chinese eventually joined the war in massive numbers, you have to ask yourself, "Why didn't the threat of nuclear war deter the Chinese?" If moving nuclear-capable bombers to Great Britain had been able to stop the Russians from escalating the crisis (as nuclear believers claim), why couldn't this same threat prevent the Chinese from joining the war? Again, this seems like a case of nuclear deterrence failing.

MIDDLE EAST WAR OF 1973

We've already talked a little about the Middle East War and the Israeli decision not to use nuclear weapons in that war. But what's interesting about this episode is the way that nuclear believers typically tell the story. In their telling, the U.S. decision to put its nuclear forces on a higher level of alert on the night of October 24, 1973, was the critical event in the war. The U.S. action came at a very dangerous point. Israeli forces had broken through west of the Nile, and they were threatening to encircle the Egyptian Third Army. The Soviet Union was preparing to fly its own airborne units into Egypt to reinforce their ally and fight Israeli forces. Putting U.S. nuclear forces on alert, according to Nixon administration officials, was intended to warn the Soviets not to send their paratroopers. And, according to nuclear believers, it worked. Raising the alert status of U.S. forces worldwide sent a message to the Soviets that caused them to eventually cancel the move. It is, of course, not clear at all that the worldwide alert had anything to do with the resolution of the crisis.

What's interesting about this event is that it is presented as the main lesson to be learned about deterrence from the 1973 war. Nuclear believers point to the U.S. alert and say, "Well, nuclear deterrence clearly worked." They apparently are ignoring, or perhaps cannot even see, a far more important "lesson" that this war teaches—a lesson with much more important implications. How is it that Israel's nuclear forces did not prevent the attack on the occupied territories to begin with?

Israel's nuclear weapons were an open secret in 1973. Their existence had been reported in the *New York Times*. Both Anwar Sadat, Egypt's leader, and Hafez al-Assad, the leader of Syria, must have been aware of the existence of Israeli nuclear forces. Neither Egypt nor Syria had nuclear weapons. How could these two nonnuclear forces have made war on a country that had nuclear weapons?

Nuclear believers argue that Egypt and Syria had limited aims in the war. They didn't want to destroy Israel; they only wanted to recapture the occupied territories. Israeli government officials, they claim, would have known that the Egypt and Syria's aims were limited. If Israeli leaders knew that Egypt and Syria had limited aims, they would know that their vital security interests were not threatened, so they would know that using nuclear weapons was not necessary. Sadat and Assad would have known the Israelis had no reason to use nuclear weapons, they argue, and so they could proceed with the war.

This seems highly unlikely. The first problem with such a suggestion is that it is notoriously difficult to know what your opponents are doing during war. They have already violated the existing state of affairs by launching a war: how can you know that their ambitions go only this far and no further? The geography of Israel compounds the problem. Israel is a very small country. At its narrowest point, it is only 9.2 miles across. If Syrian forces had broken through on the Golan Heights, they could have been in Tel Aviv in less than a day. It seems unlikely that Israeli government officials could be certain enough that Egypt and Syria's war aims were limited to risk their survival on it.

And as we've already seen, we now know that Israeli officials were not confident that Syrian forces would stay in the occupied territories. During the second day of the war, when Israeli forces were being pushed back all along the front by Syrian forces, Israeli Defense Minister Moshe Dayan apparently became convinced that the situation was getting out of hand, that the Syrian forces could well break through, and that drastic measures were called for. If Israeli government officials were seriously discussing the use of nuclear weapons at the highest level, it seems clear that they were not confident that the Syrians and Egyptians had limited war aims and would stop once they retook the occupied territories. It is simply not true that Sadat and Assad were able to be certain that Israeli leaders would not use nuclear weapons—even the Israelis themselves weren't sure.

Therefore, nuclear deterrence simply failed in the Middle East War of 1973. Egypt and Syria's leaders were not deterred by the risk of nuclear attack.

FALKLAND ISLANDS WAR OF 1982

The Falkland Islands are British possessions 7,925 miles from London. They are part of what was at one time a far-flung empire. Argentines, however, had felt that the British claim to the islands was unjustified, and a long-running dispute over the islands had festered for decades. The Argentines insisted that the islands, which lie only 414 miles from the coast of Argentina, should be called the Malvinas, that they had originally been held by Argentina, and that they should be returned to Argentine sovereignty. The British adamantly refused to do so.

So, on April 2, 1982, Argentine forces landed on the islands, successfully capturing them and immediately declaring that henceforth, they belonged to Argentina. The United Kingdom responded by sending a fleet and marines to the South Atlantic, and after sharp fighting in which a cruiser, two destroyers, a submarine, and a number of other ships, aircraft, and supply vessels were lost, along with a total of 907 lives and 2,432 wounded, the islands were recaptured and British sovereignty reestablished over them.

The problem this war presents to nuclear weapons theorists is that while the United Kingdom had nuclear weapons, the Argentines did not. And unlike some other crises involving nuclear weapons, the Argentines did not have a nuclear-armed ally whose presence might explain away the failure of deterrence. If the Argentines had a nuclear ally, then it could be argued that they felt protected by their ally's nuclear weapons—the way a friend's umbrella protects you from the rain. They might have felt confident that the United Kingdom would not use their nuclear weapons because the Argentine's nuclear ally would retaliate against them if they did. But Argentina had no such alliance.[188] So, this raises the question: how could the Argentines have attacked a nuclear-armed state and not feared retaliation? Nuclear deterrence argues that the threat of nuclear war is a frightening and powerful threat. If nuclear deterrence works, how could the Argentines have run such a risk? Again, this looks like a clear failure of nuclear deterrence.

GULF WAR OF 1990

On January 17, 1991, U.S. land, air, and naval forces and their allies attacked Iraqi forces that had invaded and occupied Kuwait. They rapidly recaptured Kuwait and destroyed most of the Iraqi forces that faced them. During these land operations, and throughout the war, there was concern that the leader of Iraq, Saddam Hussein, would order the use of chemical or biological weapons against U.S. troops (or even against Israeli civilians, who were within range of Iraqi missiles).

To forestall such a move, during a meeting with Iraqi Foreign Minister Tariq Aziz, which took place before the ground offensive started, U.S. Secretary of State James Baker handed Aziz a letter which said that if "God forbid ... chemical or biological weapons are used against our forces—the American people would demand revenge." Baker amplified the threat verbally, saying, "This is not a threat but a pledge that if there is any use of such weapons, our objective would not be only the liberation of Kuwait, but also the toppling of the present regime." Baker went on to say that the United States would use "the full measure of force" to accomplish these ends. Baker later explained that he "purposely left the impression that the use of chemical or biological agents by Iraq would invite tactical nuclear retaliation."[189] Subsequently, General Kevin Chilton, former commander of all U.S. nuclear forces, wrote in an article in Strategic Studies Quarterly that this episode was a model for how nuclear deterrence could work.[190]

However, a closer examination of the facts seems to again suggest that nuclear deterrence advocates are pointing to success and ignoring failure. It turns out that Secretary Baker's letter was not just about the use of chemical and biological weapons. It actually drew three red lines in the sand: use of chemical or biological weapons, setting Kuwaiti oil wells on fire, and "terroristic attacks" against U.S. friends and allies. Any one of those three, according to the letter, would trigger a U.S. response that would use the "full measure of force." While it is true that Iraqi forces did not use chemical or biological weapons against U.S. ground troops, they did set between 605 and 732 Kuwaiti oil wells on fire, and they did launch repeated scud missile attacks against Israeli civilians—which means that if the nonuse of chemical weapons against U.S. forces was a success for nuclear deterrence, setting the oil wells on fire and attacking Israeli civilians were nuclear deterrence failures. Obviously, this one episode doesn't determine how effective nuclear deterrence is. But one could reasonably argue from this case that nuclear deterrence only works about one-third of the time.

Not only does this episode raise questions about whether nuclear deterrence works reliably, but it exposes again the tendency of nuclear believers to shade the facts. The world they see is so strongly influenced by the lens of nuclear belief that they can take evidence that any reasonable person would say is, at best, ambiguous and could easily be interpreted as failure and claim instead that it proves what they believe.

CUBAN MISSILE CRISIS

Perhaps the clearest and most hair-raising example of a deterrence failure that could easily have ended in nuclear war but didn't comes from the Cuban Missile Crisis. President John F. Kennedy knew that if he blockaded Cuba, the crisis might spiral out of control. With so many forces so close to each other with so much tension, something might easily have gone wrong that would escalate step-by-step to worldwide devastation. But despite these concerns, Kennedy went ahead and blockaded Cuba.

In other words, nuclear deterrence—fear of nuclear war—didn't restrain Kennedy. And it turns out that concerns that things might get out of control were wholly justified.

IT IS POSSIBLE

On Saturday, October 27, 1962, at the height of the crisis, with the world on the brink of nuclear holocaust, an American U-2 spy plane flying over the North Pole malfunctioned and strayed off course—three hundred miles inside the Soviet Union. The Soviets scrambled MiGs to shoot it down. The U.S. scrambled F-102s to find it and protect it. But because it was the height of the crisis, U.S. generals had increased the alert level to DEFCON 2—one step below nuclear war—and Air Force policy dictated that the conventional air-to-air missiles on all U.S. fighters in that command be replaced with nuclear air-to-air missiles. The only weapons those fighters had as they roared toward Soviet airspace were nuclear weapons.[191]

If those two groups of fighters had found each other and fought it out, there would have been a nuclear explosion over the Soviet Union—and very likely a nuclear war. They didn't run into each other, but it wasn't the magic of nuclear deterrence that prevented nuclear war. It was luck. The fate of civilization—your fate and my fate—was decided by chance.

* * *

These incidents raise deep and serious questions about the ability of nuclear deterrence to work perfectly all the time. Remember: nuclear deterrence works most of the time, but it has to do better—it has to be perfect. The consequences of even one failure could be catastrophic. These episodes are important evidence about the reliability of nuclear deterrence. They speak to how great the ongoing risk of failure is. A realist who wanted to make a case for the long-term viability of nuclear deterrence would feel compelled to address each of these episodes in detail and analyze them at length. Yet there is no such careful, scholarly treatment of these apparent failures. There is no widely taught book by a nuclear believer titled Explaining Nuclear Deterrence Failures.[192] Nuclear "realists" haven't checked, rechecked, and dismissed these arguments; they don't seem to have looked into them carefully at all. There is a vast literature on the Cuban Missile Crisis, for example; but almost none of it deals with why the risk of nuclear war didn't deter President Kennedy from blockading Cuba. Or, as another case in point, no one has written her thesis exploring the "Falkland Islands Deterrence Failure." Open the works of the leading historian of the Cold War, John Lewis Gaddis, for example, and you will find the question of whether nuclear deterrence ever failed barely mentioned, much less carefully reviewed and discussed. Rather than delving into the fact-stream of historical events in order to investigate, nuclear believers shy away like cats from water.

This lack of real discussion about nuclear deterrence failures leaves an unmistakable impression. It makes it seem as if nuclear deterrence is not a phenomenon to be carefully and objectively explored but, rather, a faith to be defended and sustained. Genuine realists don't turn a blind eye to facts—that is a believer's sin.

Surprisingly, nuclear believers respond to this evidence by simply denying it. They claim that Berlin, the Cuban Missile Crisis, the Middle East War, and so on are not actually deterrence failures. They point to the fact that there has never been a nuclear war and say that this proves nuclear deterrence has never failed. These examples, they say, are proof that something failed but it wasn't nuclear deterrence. "The proof that deterrence has never failed is the absence of nuclear war," they say.

The flaw in this position is so clear that it is hard to believe that they are making this argument seriously. Of course, nuclear deterrence can fail without a nuclear war occurring, because nuclear deterrence is not the only thing that can prevent a nuclear war. Luck, circumstances, the unexpected actions of others, acts of nature, and even (if God is part of your belief system) divine intervention can all prevent nuclear war. Asserting that nuclear deterrence has never failed because we've all survived this long is a peculiar and, frankly, somewhat risible line of argument.

Imagine you're driving along in your car and your brakes suddenly fail. You suck in your breath, your muscles tense, and you grip the wheel with inhuman force. Even though you know the brakes have failed, your foot keeps repeatedly stomping the brake pedal to the floor. (At some level your mind refuses to believe this is happening and seems to imagine that if you just keep pumping the brakes, they will somehow magically "unfail." This is the kind of thinking human beings do in a crisis.) You're heading downhill, and the highway curves sharply to your right. The weight and speed of the car and the laws of centrifugal force are pulling you inexorably to the left, forcing you into three lanes of oncoming traffic. There's no divider and you can't stop the car from careening into the cars coming the other way. Time seems to slow. You know suddenly with a sense of finality that you're going to die. Then your eyes widen slightly as you see what looks like a small gap in the oncoming flow of traffic. At least there might be a hole in the first two lanes. Perhaps there's a chance. You swerve wildly to try to reach it, a guy in a red sports car veers out of your way, you swerve back to avoid a delivery truck, a woman in a red minivan slams on her brakes, tires smoking, and, because she fishtails sideways, makes just enough of a gap for you to squeeze through the last lane of traffic. You look up and see a telephone pole dead ahead, and some instinct prompts your hands to wrench the wheel to the right. The car rocks violently, the tires screech as you fishtail again, and then, as you reflexively jerk the wheel back the other way, there is a sound of tearing and screaming metal as the pole scrapes along the length of the side of your car and sheers off the side mirror. You finally come to a stop in the field of tall grass on the other side of the road. You've survived. It's a miracle. A whole series of accidents, unlikely maneuvers on your part and defensive maneuvers on the part of other drivers, and luck (it was lucky that the telephone pole wasn't another five feet to the right) have preserved your life.

Do you get out of the car, shake your limbs, try to catch your breath, and then say, "Well, at least I know my brakes didn't fail, because I'm not dead!"? Do you tell the policeman who arrives to write up the aftermath of your near-death experience that whatever the problem was, it wasn't your brakes? Do you argue with the repairman once the car is in the shop that the brakes don't have to be looked at because "If there wasn't a catastrophic outcome, then my brakes can't have failed"? If you do, most people will conclude that the strain of the experience has affected your brain. Of course your brakes failed. But skill and luck and other people's reactions and a very specific set of circumstances all combined to somehow avoid catastrophe. The fact that you survived is not proof that any one factor was the reason you're still alive. It's entirely possible for one safety device to fail (in this case, your brakes) but for you to still survive. No complex event is ever controlled by a single variable.

When nuclear believers say that nuclear deterrence can't have failed because there's been no nuclear war, they are making exactly the same sort of nonsensical statement as the driver of a car who says his brakes didn't fail. They are ignoring the same reality that holds for other crises. Nuclear deterrence can fail, but other factors can keep war from breaking out. Deterrence can fail, the leader can order an attack, but the commander of the rocket forces can decide on his own not to launch the attack. Deterrence can fail, the leader can order an attack, but an improbable lightning strike can knock out communications between the capital and the command center for two hours, and in the intervening time the leader regrets her choice and changes her mind. Deterrence can fail, the leader can decide to order an attack, but before he can stride into the situation room to issue the order, he has a sudden massive heart attack and dies. Deterrence can fail, the leader can decide on an attack, but before he can order it, word comes that a third nation has stepped in with a persuasive offer to mediate. There are thousands of ways for deterrence to fail and yet nuclear war still be avoided. The lack of nuclear war is no proof at all that there haven't been failures of nuclear deterrence.

The fact that nuclear believers still claim that nuclear deterrence has never failed, despite plentiful and clear-cut evidence, shows how powerful the nuclear mindset can be. It allows people to deny obvious reality. It allows them to pretend that nuclear deterrence works perfectly, when clearly it doesn't.

DETERRENCE REALITY

The reality is that the world in which nuclear believers claim to be living—the world where nuclear deterrence "has never failed"—is a fantasy world. The claim that nuclear deterrence has never failed is clear evidence that nuclear believers do not see reality, do not acknowledge reality, and are not basing nuclear policy on reality. The government officials and experts who make the policy that regularly risks your life and my life apparently can't bring themselves to be realistic about the danger involved in relying on nuclear deterrence.

*　　*　　*

A great deal has been written about the complex systems that control the nuclear weapons of the United States and Russia. The United States has an elaborate set of safeguards against accidental launch. For example, before a launch can take place, a verified launch order has to be received, a launch code has to be entered, and two missile officers have to insert two keys at the same time that are far enough apart so that a single person couldn't reach both at once. It sounds reassuring. But the reality, as always, is more human and less foolproof than it sounds.

Take the launch code. This was a matter of some fierce debate. Secretary of Defense Robert McNamara wanted it; the Air Force resisted. McNamara had been trained as a systems analyst and knew all about unexpected occurrences. The leaders of the Air Force argued that the missile launch officers were handpicked and personality tested to be reliable. They went several rounds, and the dispute got heated. In the end, McNamara won and a special digital lock was installed so that an eight-digit code had to be entered before the missiles could be launched. It was ten years before anyone figured out that the Air Force had had all the launch codes set to 00000000, which was widely known by missile launch officers, and which completely undermined this safeguard.[193]

It is hard to make any large and complex system foolproof, and through the years, the U.S. system has had numerous mishaps and problems. A forty-six-cent computer chip malfunctioned, causing the big board at the nuclear command center to show scores of incoming missiles. The problem was discovered before a counterattack was ordered.[194] One time someone accidentally put a training tape into the computer that controls the big board, causing a full-scale Soviet attack to appear on the screen.[195] Again, the problem was identified before a missile launch was ordered. A fully operational hydrogen bomb was accidentally dropped over North Carolina. Six of the seven safety features on the weapon failed, bringing the good people of North Carolina within a hair's breadth of experiencing a nuclear attack firsthand.[196] A wrench dropped while repairing equipment in a missile silo in Arkansas caused a leak of jet fuel and eventually an explosion that shot the nuclear warhead on top of the missile a good distance away into a cow pasture. The warhead did not go off. And so on. The Bulletin of the Atomic Scientists at one time compiled a list of 1,250 U.S. incidents between the years of 1950 and 1968.[197] Complex systems involving highly dangerous weapons sometimes fail.

Perhaps the most hair-raising and at the same time inspiring account of an accident that almost led to nuclear war comes from Russia. Stanislav Petrov was a lieutenant colonel in the Soviet rocket forces who worked in the command center that evaluated signals from the different Soviet systems designed to warn of an incoming nuclear attack. On the night of September 26, 1983, Petrov was working an extra shift for a friend when the system suddenly showed a missile launch from the United States heading toward the Soviet Union. After tense attempts to find out if a secondary system could confirm that the warning was false, Petrov decided, based on his own judgment, that it was an error. When a second launch appeared, despite the fact that the visual systems could neither confirm nor deny the warning, he still felt sure it was an error. But when the third, fourth, and fifth launches appeared, it became harder and harder to believe that what they were seeing was a system malfunction. Still, on his own initiative, disobeying orders and against military protocol, Petrov refused to send the report of an incoming attack to Moscow. Had he reported an attack and certified that it was genuine, Soviet leaders might well have launched a full-scale nuclear counterattack. In the event, Petrov turned out to be right. You can see an excellent movie about the event, called *The Man Who Saved the World*, which is both moving and includes Petrov himself discussing the importance of disarmament.

It is absolutely true that complex systems can fail. And it is absolutely true that there have been many "near misses" both with nuclear warning systems and with actual weapons around the globe.[198]

But the main problem is not complex systems. Complex systems can fail and have had their share of false alarms. But complex systems can be simplified. Checks and double-checks can be put in place. More and more rigorous training can be insisted on. More safeguards, more time, and more money can be expended. Ultimately, with enough resources, any system can be made safe. The problem is not the systems. The problem is not the machines. Our nightmare vision is always focused on technology. In our minds, technology is Frankenstein's monster that somehow grows too big and too complex to be controlled.

But despite what we tell ourselves, the problem is not technology. There is, in fact, another component in all these systems that can fail, a component that is included in both U.S. and Russian nuclear systems—one that is actually present in all the nuclear command and control systems of every nuclear-armed state. It is a component that not only can fail but is prone to fail. It has a history of failure. And that component is … us. We are the weakest link. The abiding dilemma, the unsolvable problem at the heart of relying on nuclear weapons for safety is the people who run the machines.

We are, after all, only human. We are flawed. And we know that we are flawed. There is no such thing as a person who never makes a mistake. Even the best of us can never match the criterion of perfection. Even saints, even the highest and holiest, are still just weak humans. Fallibility is our lot.

Human beings make mistakes. They misjudge, they assume what isn't the case, they get carried away by their emotions, and they sometimes even lose their minds and act without reason. These realities are hard to accept. In our heads, there is always a small voice that whispers that, while others may be flawed, while others may act in ways that are blameworthy, we can avoid this fate. We can be responsible in every circumstance. We can avoid the traps that others fall into. When crisis looms and pressure mounts, we will not get carried away by emotion. We can be perfect.

But of course, nothing can hold back human folly. All of us are subject to it. You are. I am. From the lowest soldier to the highest leader, everyone makes mistakes. Even those with the gravest responsibilities. It is an inescapable part of being human. This truth should be well known to Americans: it is embedded in their form of government. The founders of the United States were so aware of human fallibility that they built checks and balances into the core structure of the government they created. No individual, they believed, should be empowered to make momentous decisions without the help and counsel and sometimes restraint of others. No one can escape the reality of our nature.

For generations, people have refused to face the implications this fact—the fact of human fallibility—leads to. This is a dangerous choice. What happens when you ignore reality? What happens when you believe something so strongly that it warps your perceptions so you can't see how things really are? How does history treat nations that sin against the gods of realism? Some, of course, beat the odds. Some get off scot-free. They act badly, tempt the fates, and ultimately walk away unharmed. During World War I, French generals, for example, ignored the immovable reality of the defensive advantage that trenches confer; and nearly five percent of France's population—mostly young men—died in fruitless charges against enemies protected in trenches. The British Empire spent heavily on dreadnoughts and destroyed their national economic health. But they—the French and the British—were lucky. Their countries survived their failure to face the facts. Others were not so lucky. The Carthaginians denied their vulnerability to Rome and paid with their lives. Of the one million who lived in Carthage, only fifty-five thousand were left alive after the starvation, slaughter, and fire that the Romans brought down on them. Those left alive were all sold into slavery.

Nuclear believers refuse to confront the obvious problems with the record of nuclear deterrence. They close their eyes and repeat the phrases they have learned—"nuclear weapons keep us safe; they are the ultimate insurance"—like a catechism. The failures at Berlin, Korea, the Cuban Missile Crisis, and so on are swept under the rug. But it is not just that the past failures of nuclear deterrence are alarming. What is truly troubling is how clearly the danger that lies in the future can be seen. It is fairly easy to make a case that nuclear weapons threaten our future. A very compelling case. Actually, an irrefutable one. Nuclear deterrence will inevitably fail one day. That is not a guess. That is not a gloomy prognostication. It is an indisputable fact. It is as certain as human nature and as clear as the logic of three short sentences.

Part of the problem is that the errors we make do not occur in isolation. They are not a single action that plays out and then is gone. Like a pebble that begins a landslide, one mistake causes other mistakes in response, which cause further mistakes. As Robert Kennedy wrote in his memoir of the Cuban Missile Crisis:

> Miscalculation and misunderstanding and escalation on one side bring a counterresponse. No action is taken against a powerful adversary in a vacuum. A government or people will fail to understand this only at their great peril. For that is how wars begin—wars that no one wants, no one intends, and no one wins.[199]

Freeman Dyson, a distinguished physicist with long experience with technological systems, wrote about how unlikely it is that relying on nuclear deterrence will keep us safe.

> The concept [of Assured Destruction] is ultimately suicidal. The concept rests on the belief that, if we maintain under all circumstances the ability to do unacceptable damage to our enemies, our weapons will never be used. We all know that this idea makes sense so long as quarrels between nations are kept under control by statesmen weighing carefully the

> consequences of their actions. But who, looking at the historical record of human folly and accident which led us into the international catastrophes of the past, can believe that careful calculation and rational decision will prevail in all the crises of the future? Inevitably, if we maintain Assured Destruction as a permanent policy, there will come a time when folly and accident will surprise us again as they surprised us in 1914. And this time the guns of August will be shooting with thermonuclear warheads.[200]

The proof that nuclear deterrence must eventually fail goes like this: First, human beings are fallible. Not just the weak and ordinary. All of us. No one is perfect. Second, nuclear deterrence involves human beings. Human beings make the threats, other human beings evaluate them, and then human beings decide how to respond. Nuclear deterrence isn't a machine that runs on its own, quietly humming along in a corner. Human beings are a part of the process at every step. Third, and finally, if humans are prone to folly (and they are), and if nuclear deterrence involves human beings (and it does), then nuclear deterrence is inherently fallible. It will fail. Nuclear deterrence cannot be perfect because we cannot be perfect. So, as the years go by, nuclear deterrence will fail again and again at intervals; and one day our luck will run out and a deterrence failure will lead to catastrophic nuclear war. The logic is simple:

1. Human beings are fallible.
2. Nuclear deterrence involves human beings.
3. Therefore, nuclear deterrence is fallible.

These are the facts. They are undeniable. It will all end catastrophically one day. It is not a question of if; it is just a question of when.

Robert McNamara, secretary of defense for two presidents and one of the key participants in the Cuban Missile Crisis, emphasized this point. Looking back on his experience in that crisis, McNamara drew a frightening conclusion:

> I want to say, and this is very important. At the end, we lucked out. It was luck that prevented nuclear war. We came that close to nuclear war at the end. Rational individuals—Kennedy was rational, Khrushchev was rational, Castro was rational—rational individuals came that close to total destruction of their societies. And that danger exists today.

The major lesson of the Cuban Missile Crisis is this: The

indefinite combination of human fallibility and nuclear weapons will destroy nations.[201]

CONCLUSION

In McNamara's judgment, if we continue to rely on nuclear deterrence, eventually we will end up with nuclear war. It is unavoidable. And as long as we have nuclear arsenals of any kind, that will be the case. The only way to prevent nuclear war is to eliminate nuclear weapons.

If nuclear deterrence is not perfect, if it cannot work every time, then it is not safe to rely on nuclear weapons—because in order to be safe, nuclear deterrence has to work all of the time. It has to be perfect, not probabilistic. The consequences of nuclear deterrence failure could be mortal for the United States, for Russia, for China—for any of the nuclear-armed states. Deterrence failure could lead to catastrophic all-out nuclear war, and all-out nuclear war would likely mean at least two hundred million people dead and civilization wrecked for centuries. Such an outcome is simply unacceptable. If even the smallest risk of all-out nuclear war exists, then relying on nuclear deterrence is not an option—no matter what the benefits.

Advocates of nuclear weapons believe that fear of nuclear war—nuclear deterrence, in other words—can prevent us from being foolish. Nuclear deterrence, they think, can hold off our inbred desire for all-out war. But there is another power that is also at work here: the power of the wrath of war, that savage desire for revenge and victory, the overwhelming urge to punish the enemy, that so often grips humanity. Decades can go by—even a hundred years—without it overpowering us. But it is a savagery that only sleeps. It stretches back into our deepest past. Here is Homer's description from Book XVIII of *The Iliad* of the wrath that lives deep in the human heart, the wrath —

> that makes a man go mad for all his
>   goodness of reason,
> That rage that rises within and swirls like
>   smoke in the heart and becomes in our madness
>   a thing more sweet than the dripping of honey.[202]

Look at the news. Think about the past few years. Look at the war in Ukraine. The tide of violence and extremism is rising. You can feel it. Everywhere, on all sides, there is growing anger and intolerance. Anti-immigrant prejudice, religious fundamentalism, and political extremism exist on all sides. There is increasing zeal and hatred in Europe, in Asia, in the United States. Those emotions are a warning sign. If nuclear weapons are near at hand when the wrath of war next overwhelms us, they will be used. These are pessimistic thoughts. But realism is pessimistic. No one who has read the history of human beings closely can doubt the power of our deep-rooted self-destructiveness.[203]

The government officials in charge in the nuclear-armed states can't bring themselves to look reality in the face. They have been tempted to believe in the seductive power of the monster-god. They have talked themselves into a world in which mistakes can be avoided, in which danger can be safely ignored. This is deeply concerning because this is not simply their problem. After all, our fate is in their hands. If they stumble into nuclear war, we will pay with our lives. Do you believe the government officials who confidently tell you that everything is perfectly safe? Do you believe that they can avoid mistakes with nuclear weapons in perpetuity?

If we do nothing, human folly will certainly overwhelm nuclear deterrence, anger will drive someone to use nuclear weapons, and escalation will cause that use to spiral into all-out war. Human fallibility mixed with nuclear weapons is a dangerous cocktail. We live in a swiftly darkening landscape. There is still hope. But we must act quickly, with determination, courage, and—most of all—with a clear-eyed view of reality.

[184] Several of the options discussed by the American policymakers were rejected because they posed "the grave danger of war." Herken, *The Winning Weapon*, p. 258.

[185] The bombers were not, in fact, nuclear-capable. Because of the size of the nuclear bombs in the U.S. arsenal at the time, bombers that were going to drop the bombs had to have specially enlarged bomb bay doors. The bombers moved to Great Britain did not have these.

[186] Roger Dingman, "Atomic Diplomacy during the Korean War," *International Security*, 13, no. 3 (Winter 1988–1989), pp. 54–55.

[187] Gaddis, *The Long Peace*, p. 110.

[188] Technically this is not true. The Monroe Doctrine and the Roosevelt Corollary, although not a formal treaty, are generally taken as a commitment by the United States to protect any Latin American country attacked from outside of the hemisphere. However, the United Kingdom's long-standing and close relationship with the United States, resolutions in the United Nations, the Commonwealth of Nations, both houses of the U.S. Congress, and other institutions condemning Argentina's attack, and the sometimes-difficult relationship between the United States and Argentina made it highly unlikely that these commitments would be carried out against the United Kingdom.

[189] See Ken Berry, Patricia Lewis, Benoit Pélopidas, Nikolai Sokov, and Ward Wilson, "Delegitimizing Nuclear Weapons: Examining the Validity of Nuclear Deterrence," pp. 31-32; and Ward Wilson, *Five Myths about Nuclear Weapons*, (New York: Houghton Mifflin Harcourt, 2013) pp. 83-84.

[190] Kevin Chilton and Greg Weaver, "Waging Deterrence in the Twenty-First Century," *Strategic Studies Quarterly* (Spring 2009), p. 33.

[191] Dobbs, *One Minute to Midnight*, p. 264.

[192] For a useful exception, see John Orme's work. John Orme, "Deterrence Failures: A Second Look," *International Security*, 11, no. 4 (Spring 1987), pp. 96–124.

[193] Bruce Blair, "Keeping Presidents in the Dark," Bruce Blair's Nuclear Column, February 11, 2004, http://web.archive.org/web/20120511191600/http://www.cdi.org/blair/permissive-action-links.cfm (accessed May 25, 2023).

[194] Eric Schlosser, *Command and Control: Nuclear Weapons, The Damascus Accident, and the Illusion of Safety* (New York: The Penguin Press, 2013), pp. 367–368.

[195] Schlosser, *Command and Control*, pp. 365–366.

[196] "New Details on the 1961 Goldsboro Nuclear Accident" The National Security Archive, June 9, 2014, https://nsarchive2.gwu.edu/nukevault/ebb475/ (accessed May 25, 2023).

[197] Chuck Hanson, "Review of The Hidden Cost of Deterrence: Nuclear Weapons Accidents," *Bulletin of the Atomic Scientists*, 46 (October 1990), p. 43.

[198] An excellent summary of these accidents and the case for disarmament based on them is made in Eric Schlosser, *Command and Control*.

[199] Robert Kennedy, *Thirteen Days: A Memoir of the Cuban Missile Crisis* (New York: W. W. Norton & Company, 1969), p. 125.

[200] Dyson, Aron, and Robinson, *Values at War*, p. 29.

201 *The Fog of War*, directed by Errol Morris (Sony Pictures Classics, 2003), 14:41–15:49.
202 Quoted in Robert F. Kennedy, *To Seek a New World* (New York: Bantam Books, 1967), p. 233.
203 See, for example, Matthew White, *Atrocities: The 100 Deadliest Episodes in Human History* (New York: Norton, 2012).

12. ELIMINATION

*It is possible because when people learn the facts,
they want to get rid of nuclear weapons.*

Nuclear weapons are the most catastrophic problem we face as a civilization. It is true that climate change could have more far-reaching effects over the long run, but climate change is necessarily gradual, whereas nuclear war could kill millions, destroy untold wealth, and devastate civilization in a matter of hours.

And the spike in anxiety about nuclear weapons in recent years reflects a growing sense of foreboding that nuclear war is unavoidable. Arsenals are now growing or being upgraded in every nuclear-armed state (we appear to be in a second nuclear arms race). Recently one of the world's most important powers—China—which for a long time had a rather limited arsenal, has increased the number of nuclear weapons it possesses substantially. And over the past few years, threats to use nuclear weapons have become ever more frequent and ominous.

There is little evidence that people feel any hope that these increasing dangers can be managed. After seventy-five years of failure, most people have lost faith that any escape route exists.

But the truth is, despite this fatalism, the elimination of nuclear weapons is within our grasp. The obstacles are mostly symbolic and emotional, not practical or realistic. For the most part, we are downhearted because these obstacles have been inflated by unwarranted pessimism rather than by any realistic assessment of the situation. Despite the fact that most people cannot imagine how these weapons could possibly be eliminated, there is an obvious and practical path away from danger.

IT IS POSSIBLE

To eliminate nuclear weapons, you need only to argue forcefully that they are both dangerous and not very useful—in other words that they are obsolete—and do so effectively enough that a consensus develops around that view. If nuclear weapons were considered obsolete, their permanent elimination would be achievable. Obsolescence is the key criteria for abandoning any type of technology. No one keeps tools or technology that are obsolete. If a case can be made that nuclear weapons are obsolete, then abandoning them is the only logical next step. And once that consensus takes hold, the danger that some future leader would restart the arms race in nuclear weapons would be negligible: no one races to build weapons that are obsolete. (Although stringent safeguards would nonetheless be prudent.)

"But," advocates for these weapons will say, "nuclear weapons are obviously not obsolete, they have been the technology that people have relied on for security for the better part of a century and that some 4.2 billion people rely on today." Despite these assertions, however, their lack of utility and the danger they pose has been obvious all along, hiding in plain sight. People failed to see these limitations because they were obscured by strong emotion, the growth of a false mythology, and an almost willful determination to distort past evidence. And these beliefs were topped off with the warning that nuclear weapons will only be declared obsolete when something bigger and more horrifyingly destructive is invented.

The emergence of even more horrible technology, however, is not the only way that a weapon can become obsolete. While it is true that obsolescence most often occurs when existing technology is replaced by newer, better technology, there is also a second way for obsolescence to occur. It is rarer, but it does happen. Technology can be abandoned if people eventually come to realize that their initial fascination with it was misplaced. Sometimes a new technology seems remarkable and highly desirable, it is adopted and used, but over time people eventually decide that it was not, after all, as useful as they at first thought it was.

Take the large cars built in the 1990s that were modeled on military Humvees, called Hummers. People initially loved these cars. But over time it became clear that they were difficult to park (because they didn't fit in most parking spaces), that their size made the chances of sideswiping another car or a telephone pole greater, and that there were many places they simply could not go. Older cities, because of their narrow streets, were simply not accessible in a Hummer. You had to get out and walk. Not to mention that when the price of gas went up, Hummers became painfully expensive.

It took some time for this reality to sink in. But eventually Hummers were abandoned. Not because something bigger came along, but because people realized their mistake. This same form of obsolescence came to the Paris Gun, to Ptolemy's giant warship, and to other types of technology over the centuries.

So although technology typically gets abandoned when better technology comes along, sometimes it is abandoned when people realize their initial enthusiasm was misplaced. Nuclear weapons can be eliminated if we can show that the initial exaggerated beliefs about them, propagated during the early years of the Cold War, were wrong.

And we can dispel those early beliefs and reassert the importance of reality. Because where survival is at stake, reality is king. People may hold on to congenial illusions when there is little danger. But the threat of extinction concentrates the mind. No one suffers foolish illusions when lives are at stake. Reality is always a winning argument when the issue is survival.

* * *

So this is our road map. First, argue forcefully that nuclear weapons are obsolete (and have been from the beginning). They have virtually no military utility and are catastrophically dangerous. Then drive home these facts in every nuclear-armed state and every state that relies on another state's nuclear weapons for protection. Finally, once the reality sinks in and consensus takes hold, eliminate nuclear weapons from all countries at the same time.

This approach—which has never been attempted or even explored—is entirely based on a pragmatic assessment of the weaknesses and strengths of human nature. It aligns with the facts. And it gives us, for the first time in a long time, a reason to hope.

MOVEMENT

Once the general strategy is decided on—a pragmatic argument that nuclear weapons are too dangerous and have so little utility that the only practical course is to abandon them—we must turn to the tactical questions. And the tactical difficulties are considerable.

Today the nine nuclear-armed states and their allies insist that nuclear weapons will always exist, that constant and ongoing threats to slaughter millions of civilians are a necessary part of everyday life, and that nuclear weapons will always be the final arbiter of world affairs. Policy debates and decisions in those nine countries are dominated by small, siloed-off elites who believe that only they understand the issues at hand, who dismiss the ideas and objections of anyone outside their circle, and who therefore cannot be engaged in reasoned discussion.

You have seen in the last eleven chapters that the central tenets of these small policymaking groups are deeply flawed, that nuclear weapons policy is built on a foundation of assumptions so obviously unrealistic that they can only fairly be called myths. Those unrealistic ideas constantly endanger the lives of the billions of people whose governments insist on relying on nuclear weapons for safety.

So, what is to be done? If these policymakers refuse to hear objections or doubts, if there's no way to get them to hear, how is it possible to get a hearing for the idea that these weapons are obsolete, much less build a consensus around that view? This obstacle is serious. It is, literally, a matter of life and death. Should we consider extreme measures? The use of force, for example? Continuing on the way we are going will possibly lead to catastrophe for millions of people and a large part of civilization. Should we advocate that the non-nuclear-armed states arm themselves and attempt to conquer the nuclear-armed states? Is it necessary to foment civil wars in the nuclear-armed states? What is to be done?

Violence would be counterproductive and dangerous. Any time a nuclear-armed state is involved in conflict, even if it is civil war, rising desperation makes it possible that someone will resort to nuclear weapons. So, violence is off the table. And in any case, even though the current impasse seems difficult, there is a way around it.

The policymaking elites that stand in the way can be outmaneuvered. There is a way around those elites that is rarely used but that does work. Recall that the policymaking elites are not the last word. They report to, and must abide by the decisions of, political leaders—presidents, prime ministers, members of parliaments and congresses. And those political leaders have overruled their policy advisors in the past. Mass movements—large numbers of ordinary people insisting that policy change—can alter the course a nation takes, even when the matter is as serious as nuclear weapons policy.

In the early 1980s, the nuclear arms race seemed beyond control. It had been barreling ahead at a breakneck pace for thirty years and nuclear arsenals had grown alarmingly large. From a handful of weapons in the 1940s, the world's arsenals had mushroomed to something like seventy thousand weapons by the early 1980s. There were so many weapons that planners struggled to find uses for all of them. There was one radar array in Moscow that famously had sixty-two different nuclear weapons targeted at it. But still, nuclear weapons policymakers insisted that a "clear and present danger" existed and that more weapons must be built, more weapons must be deployed in more places around the world. Sitting on mountains of weapons, they still wanted more. They were in the grasp of what can only be described as a kind of collective madness.

But in the United States and in Europe, masses of people joined together and demanded a halt to the nuclear arms race. Millions of signatures on petitions were collected; gigantic rallies were held. In the United States, somewhere between 750,000 and a million people marched in New York City—at the time, the largest protest the nation had ever seen. In Europe equally large and record-breaking protests occurred. Initiatives and referendums insisting that the arms race be halted were placed on ballots, won resoundingly, and demonstrated that ordinary people would not support any further growth in nuclear arsenals.

Political leaders heard those ordinary people's voices and changed course. They overruled their policymaking elites and altered the direction of nuclear weapons history. U.S. President Reagan met with Soviet President Gorbachev, and together they agreed to reduce their nuclear arsenals by almost half. Over time the reductions continued, and today there are only an estimated thirteen thousand nuclear weapons in the world. What this history demonstrates is that there is a way to deal with policymaking elites. When they do not listen, you can bypass them with a mass movement. If enough people speak, leaders will listen.

So, another mass movement against nuclear weapons is necessary. But this time, we are going to take aim not at reducing the size of the arsenals, not at freezing the arms race, but at ending this ominous danger once and for all. Rather than telling political leaders that they must stop the arms race or even reduce the world's arsenals, we will insist that they do away with these dangerous and useless weapons altogether. No more half-measures. It turns out that we have in our hands the tools we need to put an end to nuclear weapons. They have been there all along, but they were hidden by false claims and high emotions. Let us use them.

COMBATING SYMBOLS

But how does one construct a mass movement against nuclear weapons? As we have seen, it is their symbolic power that drives their appeal, not any military reality. So, what we need is a kind of rebellion against dangerous fantasies. Not a violent uprising, not a civil war, but a slashing rhetorical attack on the beliefs that have put us into such danger for so long. And how does one deflate symbols? With realism. We need to wield realism like a sword, puncturing the exaggerated beliefs of nuclear weapons advocates.

IT IS POSSIBLE

Because letting the air out of a symbol is an unfamiliar task for most of us, some may fear that the exercise will be difficult. The seventy-odd years of losing the debate over nuclear weapons have left many of those who argue against nuclear weapons feeling downhearted and unsure of their own strength. And for decades, the advocates for nuclear weapons have reinforced the notion that they have the upper hand in any debate. They have cast us as doomed idealists, fighting a desperate rear-guard action against an enemy that, in our heart of hearts, we know we can never defeat. But that is their story, not ours.

We are actually quite strong. We have been sleeping for the better part of a century, but we are waking from our long, spellbound slumber. For sixty years we have mumbled objections in our sleep—objections that were ignored. But now, yawning and stretching, standing by the bed looking about, we find that there is a weapon near at hand that will make us invincible. Surprised to find it there, we lift it and swing it whistling through the air. It feels good, solid and ready in our hand. That weapon is reality.

One of the truths of human nature is that myths cannot stand up to reality. Start telling someone unwelcome truths and watch what they do. They will desperately try to keep you from speaking. Only by hiding from the reality, by stuffing their fingers in their ears and repeating over and over again, "I don't hear you, I don't hear you," can people who believe in myths withstand reality's force. Their childish unwillingness to hear contrary arguments is essential to maintaining their beliefs because myths die when they are exposed to reality. The power of the truth is so irresistible that shielding your senses from it is the only way you can continue to hold onto your myths. The reason we can't convince children to continue to believe in Santa Claus above a certain age is that truth cannot be ignored, even for children who wish it isn't so.

True, there are tougher myths than Santa Claus. The ones that adults tell one another—stories about the origin of their country, beliefs about the superiority of one group over another, conclusions about the meaning of a long-ago war—can be harder to shake. And myths like these can sometimes hold a small, determined group spellbound in their grasp. But they are only held if they stay cocooned away from the facts, only if they live in an information bubble that protects them from facing the truth. Reality is impartial and remorseless. In a fair fight, reality never loses.

Realistic arguments are so powerful because of the very deeply rooted human desire to survive. It is, in our species, an overwhelming imperative. And people know from experience that foolishness does not protect their survival in the same way that careful attention to reality can. The only way to ensure survival is to face the world as it is, to see the opportunities and the risks objectively. Fantasy cannot safeguard us. Ultimately, reality always wins.

Our strength lies in the fact that we have looked more closely at reality than the advocates of nuclear weapons have. It has not always been pleasant, but we have done the work to see through their wishful thinking and hope-filled beliefs. That makes us far more formidable than any group that has ever sought to take on these weapons before.

What we need is a realists' revolt, a widespread movement of ordinary people who refuse to sit by quietly while "experts" tell them what's real and what's not. We need a campaign that puts facts in place of fantasies. We need people who are more concerned with common sense than esoteric theories. We need ordinary people who are willing to make a compelling case that nuclear weapons can be eliminated and who will do the work to make those arguments common knowledge.

MY COUNTRY

So, how would you build a national movement? There are some things that are specific to mounting a grassroots campaign to eliminate nuclear weapons and some things that are general to all grassroots campaigns. In general, a campaign to build a mass movement involves attracting supporters, repeatedly drawing attention to the issue, and ultimately using the strength of your organization to convince political leaders to change policy. In the early stages, organizing often includes going door to door, social media posts, setting up a booth at a public park or event, giving publicized talks, and so on. At the same time, you draw attention to the issue by marches, protests, getting covered by local media, and online work. Often you measure your support in the wider community with petition drives, public rallies, and referendums at election time. Different areas require different approaches. What works in one city, region, or country may not work in another. But the same general template applies. You can learn about the things that are common to all issue campaigns from any number of useful and well-argued books. You can find a list of some in the appendix to this book.

When you mount a campaign about nuclear weapons, there are two specific things you should keep in mind. First, don't try to influence the policymaking elite. You can't convince them. You can debate against experts (as a way of gathering supporters), but don't waste your time imagining that you will persuade them. The nuclear weapons elite is well practiced at engaging people with promises of influence and then telling them they have to talk using insider's jargon, have to think using insider's theories, have to argue accepting insider's assumptions, and first thing you know, you are trapped inside the insider's worldview—a mindset that makes it impossible to convincingly argue against nuclear weapons. And no matter how far you go to try to talk to them on their terms, they will never quite let you have a meaningful say in the actual policy discussion. They've spent seventy years subverting thoughtful and well-meaning groups. They know how to do it.

The second difference between a grassroots campaign about nuclear weapons and other grassroots efforts is that in a campaign about nuclear weapons, you have to break the taboo. As discussed earlier, there is a taboo around talking about nuclear weapons issues. Every day people actively try to put the subject out of their minds—they've been told there's nothing that can be done about the issue. They feel uncomfortable about the issue, and those emotions make it harder to talk to them. You'll have to overcome this initial reluctance. I've found that the key is finding a way to tell them that something actually can be done. But however you get past the taboo, getting people talking about nuclear weapons is essential.

The ultimate goal of a national grassroots campaign is a statement by that country's political leaders that says that nuclear weapons are obsolete. Remember, technology goes away when it is no longer useful. So, an admission that a piece of technology (like a nuclear weapon) is obsolete is tantamount to getting rid of that technology altogether.

A national referendum that votes that nuclear weapons are obsolete, a statement by the president or prime minister, or a vote by parliament stating that nuclear weapons are obsolete all work to commit the country to elimination. So, that is the first important goal of a national campaign.

And by the way, it is not the goal to get the country to remove nuclear weapons from its arsenal. That is not the pathway to worldwide elimination. I know I said this earlier, but let me just say it again: no nation is going to lay down its nuclear weapons unilaterally. It is a simple fact (if a strange one) that many people have somehow gotten the idea fixed in their heads that getting rid of nuclear weapons will involve the United States or some other country unilaterally surrendering its weapons and then hoping against hope that this noble act of idealism will inspire its adversaries to become good and kindhearted people who will then lay down their weapons as well. Why people have gotten this peculiar notion into their heads, I can't tell you. But they have. Maybe it is a "straw man" argument that nuclear weapons advocates created to confirm their believe that all nuclear weapons opponents are dreamy idealists. At any rate, there is very little to say about this plan. It is a foolish idea that might work one time in a thousand, but no one I know of is seriously advocating it.

But since this nonsensical notion is very widespread, let me be very clear. Are you listening? No country is going to lay down a single weapon unless everyone else does, too. Any sensible treaty calling for the elimination of nuclear weapons will require that all nations lay down their nuclear weapons at the same time. So, let's not have any more of this "unilateral disarmament" stuff. I don't want anyone to claim they've read this book and then stand up at a public event and ask, "But, Ward, if we lay down our nuclear weapons unilaterally, won't the Russians…?" Believe me, if you try it, I'll happily remind you of these two paragraphs.

CAMPAIGN FABLES

How to work at the grassroots will be familiar to political organizers. It's a playbook that has been used again and again. But most people are unfamiliar with that playbook. It's hard to picture something that you've never done before, so let me give you some examples.

REBECCA

Rebecca met Brody at an independent film viewing at New York University one night. Afterward, they got to talking, and over a beer she happened to mention the darkness that nuclear weapons seem to bring the world. She noticed when Brody nodded quietly but emphatically, and they talked for a while about nuclear weapons—shared stories about when they first heard about nuclear weapons, talked about how stupid the weapons seemed, and even admitted their fears that things might get out of control. Rebecca said that she'd heard about this exciting new idea that nuclear weapons were basically all a myth, and they talked about that over a second beer. Eventually, they decided to get a group of people together and watch a few videos about the topic.

The group meeting went well, the group kept getting together, and by the time midterms came around, there were nine people who were regularly coming to their every-other-week meetings. Someone suggested they really ought to do something, and Rebecca said why not start a petition to get the university to divest from companies that are part of the vast enterprise that manufactures nuclear weapons.

They found "Don't Bank on the Bomb" online, and within two months, they had seventy-four percent of the student body signed on to a petition, they were having regular candlelight vigils that were getting covered in the media, and even some folks from outside the university were coming to their meetings. That May, the trustees voted to begin a process of divesting the university from companies that make parts for nuclear weapons.

TAMIQUA

Tamiqua read an article in her church bulletin about the effect that constantly living under the threat of nuclear war was having on children. The article cited the widespread insecurity and uncertainty that young people—especially boys—were suffering, along with other signs of emotional trouble. One day, Tamiqua and her neighbor (who also had a fourteen-year-old son) were talking, and Tamiqua told her about the article. That led to a series of conversations about their worries for their sons and for some reason they kept circling back to the topic of nuclear weapons.

Tamiqua mentioned the conversations to her pastor, who organized a few other members of the church to meet and discuss the issue. Tamiqua invited her neighbor, who also invited a cousin of hers. The meeting led to more discussions. People had never thought about the idea that living under the threat of nuclear war all the time might affect people's sense of hope, but they had to admit it made sense. They started reading and sharing articles, learning more and more; and when a local women's organization had a march, eleven of the members of the church group marched along, carrying signs about nuclear war. A local reporter noticed the women and thought the concern about nuclear weapons was interesting and different, interviewed them for local TV, and soon the meetings had twenty or so regular attendees.

That spring, Tamiqua suggested the group lobby their local town council to vote to endorse the Treaty on the Prohibition of Nuclear Weapons (TPNW). They held an organizing meeting, Tamiqua spoke from her heart, a college professor who had started attending the meetings answered some of the objections people raised, and Tamiqua's pastor gave an impassioned denunciation of the immorality of nuclear weapons. The group started attracting more supporters—even a few people from the suburbs nearby. They worked hard to gather signatures and go to the town council meetings. And at the regular town council meeting three months later, the council voted five to two to endorse the TPNW, making their town the first in their state to take a stand.

That success led the group to think bigger. They started to look at other towns nearby and even their state legislature.

MR. KIM

Kim Chung-Hee had retired nearly a decade ago and loved spending his afternoons at a local coffee shop chatting with friends. One day, he and another retired gentleman, one he hadn't met before, fell to talking about the problems in Korea and especially the dangers posed by Kim Jong Un's nuclear weapons. The other man, whose name was Barry, was a retired lawyer and had just finished reading a book that said nuclear weapons were symbols more than reality. He talked with Mr. Kim about the book and its arguments for an hour or more. Thinking back about it, Mr. Kim decided it was one of the most satisfying afternoons he'd spent in a long time.

Mr. Kim bought the book and read it, absorbing the ideas in it and, in his quiet way, getting more and more excited. One Sunday, when his sister's grandchildren were visiting, he got to talking in the study with Martin, the grandson who was attending Yale. Martin listened, enthralled at Mr. Kim's arguments. Seeing his enthusiasm, Mr. Kim suggested that Martin write a letter to his college newspaper about the uselessness of nuclear weapons. "No," Martin said. "We should make a video!"

A week later, Martin called Mr. Kim one evening and they talked and talked about what sorts of videos they could make and what it might cost. Martin said four of his friends were interested in making a series of videos and putting them on something called YouTube.

Later that evening, Mr. Kim dug into his desk and looked at his retirement fund. Then he did some calculations and decided he could afford to put some money toward a good cause. The next day, he called Martin and told him he'd buy the equipment Martin and his friends needed. But they would have to list him as "Producer," Mr. Kim jokingly said.

The videos that Martin and his friends made got thousands of views in the first few days. After Martin showed him how to do it, Mr. Kim would log on to YouTube every few days to see how they were doing. Soon the videos had gotten hundreds of thousands of views. And each week, Martin and his friends added another video with a new set of arguments. By the time the seventh video was on the internet, all the videos had gotten more than seven hundred thousand views—and one had more than a million. Martin said they were going to make a video calling for everyone to work to put a referendum on the ballots for the November elections in their state. The referendum would be on declaring that nuclear weapons were obsolete. The videos were later credited with getting referendums passed in five states nationwide. And each video's credits ended with the words "Produced by Kim Chung-Hee."

IT IS POSSIBLE

There are many ways to persuade people, many different paths to creating consensus and political power. These fables are just examples. Here is a short list of other ways you could get involved.

- Educate yourself. The more you know about the history of nuclear weapons, the more ammunition you have for building realist arguments. You'd be amazed what a few well-chosen facts can do to win an argument or persuade someone. Then educate others. Start a reading circle. Try practice debates. Get comfortable with the arguments.

- Break the taboo. Perhaps the most important step is the simplest: Bring up nuclear weapons in conversation with your friends and relatives. Tell them you believe it is possible to eliminate nuclear weapons and that there are sound realist arguments in favor.

- Get organized. Human beings are social creatures. We crave the comfort of groups. Be a group maker. Set up a regular time, find a place where you can meet, and then recruit people to come. Or join a national group. Or make your own group and join a national group. There is a list of national organizations in the appendix to this book. See which one appeals to you and sign up. Think about volunteering or donating money.

- Talk to your elected officials. They may not agree with you. They may try to avoid the subject. But they will eventually either listen to you or get voted out. They are the ones who can override the policymaking elite. So, work on them. Don't expect immediate results but talk to them with facts. Start with local officials and work your way up. (Tip: Handwritten letters are more likely to get responses.) And vote.

- Start a newsletter, set up a website, post a series of short videos on Instagram, draw cartoons, write a song, paint pictures, stand on a soap box, make your own YouTube lecture—put your ideas about these weapons out there. (Tip: Don't be afraid to be funny. The philosopher Hannah Arendt tells us that the best way to undermine authority is with ridicule. The awe we feel toward nuclear weapons is fully half the problem. Undermine it by showing how ridiculous the weapons are.)

Politics is about persuasion, and there are many different ways to persuade. Be creative. Show you care. You can do it.

One thing to keep in mind is that changing minds can happen much faster than you expect. That is the whole point of books like *The Tipping Point*.[204] People seem to hold certain latent ideas and beliefs in their minds.

Those ideas and beliefs incubate under the surface in a way that we don't really understand. But when the moment is right, when the stars align, a change of heart and mind can sweep across large numbers of people like an avalanche. Seemingly out of nowhere, a tiny event pushes people over their tipping point—and then big changes come fast. Our job is to kick the rock that starts that avalanche.

We know from surveys that the great majority of people do not like nuclear weapons and would get rid of them if they could. So, it appears their hearts are convinced. It's just their brains that are waiting for the right argument, the right video, the right set of conditions to show them a pathway to elimination. People are waiting for someone to knock on their door and explain in everyday language that nuclear weapons are not such great weapons and that it makes sense—it's just plain common sense, actually—to get rid of them.

THE WORLD

Once one country is convinced that nuclear weapons are obsolete, how does that belief spread to the other eight nuclear-armed states? Obviously, all the nuclear-armed states have rationales for keeping nuclear weapons. Small countries like Israel and North Korea think that nuclear weapons give them the power of big countries and ensure their survival. Countries that were once more powerful like Russia, France, and the United Kingdom believe that nuclear weapons allow them to retain their former grandeur. Countries next to larger, antagonistic states, like Pakistan, think that nuclear weapons equalize the regional balance of power. India and China see them as symbols that ratify their new status as great powers. But despite these rationales, a campaign to eliminate nuclear weapons will be easier than you might imagine.

First, the justification for keeping nuclear weapons is already fragile. Think about the arguments for keeping nuclear weapons and the painfully obvious commonsense contradictions embedded in them. They lead to embarrassing questions. "Wait. In order to be safe, we have to be constantly vulnerable to total annihilation? Really?" Or "We have to keep these incredibly expensive weapons and constantly improve them because … we never want to use them?" The fact that policymaking elites have to hold themselves in such careful isolation is also something of a red flag. If their ideas were truly robust, wouldn't they be willing to discuss them in any forum?

And it's worth mentioning that achieving a worldwide consensus would only involve convincing the nine nuclear-armed states (and perhaps some of their allies). Remember that 122 countries already voted for the Treaty on the Prohibition of Nuclear Weapons in the United Nations. As I write this, ninety-five countries have signed the treaty and sixty-eight of those countries have already ratified it. (And more are on the way.) The treaty entered into force in January 2021 when fifty states had ratified it, and for those states that signed on, it is now international law. According to the Nuclear Weapons Ban Monitor, 48.2 percent of the world's countries have already either renounced nuclear weapons or support renouncing them, while only 21.8 percent of the world's countries support keeping nuclear weapons.[205] So, there is already a majority of countries in favor of eliminating nuclear weapons. Any effort by a nuclear-armed state to argue that nuclear weapons were obsolete would immediately be endorsed by a chorus of support from non-nuclear-armed states.

Even current allies of the nuclear-armed states might throw their support behind the conclusion that nuclear weapons are obsolete. Surveys show that in a number of countries that are "protected" by nuclear weapons—in Europe, Korea, Japan, and elsewhere—majorities of ordinary people (sometimes as much as seventy percent) are opposed to nuclear weapons and would get rid of them if they could just find a way to do it safely.[206] So, much of the world is already prepared, emotionally, to eliminate nuclear weapons. All they are waiting for is someone who can explain to them why it is reasonable to do it.

Once a national leader commits to eliminating nuclear weapons and policy changes, that country's political leaders and diplomats will reach out to other governments to persuade them to join the process. (If they don't, then ordinary citizens have to hold their feet to the fire.) And citizen groups will also spread the word to citizens in other countries, encouraging them to build a mass movement of their own. If the effort were led by the United States, for example, which invented nuclear weapons and promoted them for years as the currency of power, the dramatic quality of the about-face would give the claim even greater power.

And once some of the most powerful countries are engaged, they can lean on their allies. The United States will press Israel, China will convince North Korea and Pakistan, and in the end, Russia will not be able to hold out against the entire world.

Perhaps such a world campaign would be the result of one brave government declaring nuclear weapons obsolete. Perhaps it would be the result of ordinary citizens voting out legislators who support nuclear weapons in large numbers. The belief in the uselessness of nuclear weapons might emerge in one country and spread outward. Or it might bubble up in a series of countries—spreading upward from the grassroots in many places at once. Many pathways lead to the elimination of nuclear weapons.

OBJECTIONS

People will ask, "But if you eliminate nuclear weapons, doesn't that mean that they'll be replaced by something bigger, more destructive, something more horrible?" There is an assumption hidden in this question, an intuition that comes from everyday life. If you have a tiny little hammer that is so light it takes ten strokes just to drive a single nail, then you naturally want to replace it with something bigger. In this everyday example, bigger is better. But there is a point at which this stops being true. I once found a picture online of a guy who had built a hammer that had the normal proportions of a hammer but that was twelve feet long. It was, obviously, useless. To grasp the handle, you'd have to wrap both arms around it—and you'd never be able to lift it, much less use it to drive a nail. If you did, you'd drive the nail too deep and leave a dent in the wood the size of a plate. Bigger is not always better.

The weapons that replace nuclear weapons won't be bigger weapons; they will be more effective weapons. The criterion for abandonment is utility, not size. Nuclear weapons haven't been used for seventy-odd years because they are too big. What replaces them will be more intelligent, more discriminate weapons, not bigger ones. They will be tiny flying weapons that swarm the enemy. They will be David, in other words, not a new Goliath even bigger than the original Goliath.

The other objection people invariably raise when the subject of elimination comes up is this: "But what happens if the world eliminates its nuclear weapons and then some unscrupulous dictator—one who doesn't care about killing civilians—builds an arsenal in secret?" Again, the question is based on a false premise. It assumes that even a handful of nuclear weapons would be decisive. It assumes the old Cold-War idea that nuclear weapons are the ultimate weapon able to win wars with just two uses. But as this book has argued, that assumption is not supported by the facts.

Think about it. First, it's very hard to build an arsenal of nuclear weapons and keep it a secret. Uranium and plutonium, because they give off radiation, constantly signal their presence in a way that makes them remarkably easy to detect. A cheater can go to great lengths to try to erase the evidence of work on nuclear weapons—they can even bulldoze over a spot where nuclear weapons work has been going on—but there will still be detectable traces in the soil years later. And the work required to enrich uranium requires a lot of machinery, and it is hard to hide, because it is such an extensive manufacturing process. The consensus among the knowledgeable physicists I've spoken to is that you might be able to build a handful of nuclear weapons in secret, but building a large arsenal undetected is virtually impossible.

Second, even if you built, for example, thirty nuclear weapons, that wouldn't be nearly enough. You're going to be fighting the whole world. The leader who suddenly unveils a nuclear arsenal and starts to make demands will immediately be branded "another Hitler," and nations will take remarkable risks to stop him (or her). Even ten times thirty weapons wouldn't be enough. If, for example, some nation built an arsenal of three hundred weapons in secret, would they use one hundred against Russia, one hundred against the United States, and one hundred against China? That would leave none for NATO, or Japan, or India, and so on. Japan had sixty-eight cities bombed and still carried on fighting during World War II. And you would have to do more than bomb cities. You'd want to use some weapons against approaching ships and concentrations of enemy troops, not to mention the nuclear production facilities of those nations. Remember that such a war would have to be won quickly because the former nuclear-armed states would immediately start reconstituting their own nuclear weapons arsenals, perhaps as quickly as three to six months. Against a determined world, there simply isn't any way to turn a limited nuclear weapons arsenal into a winning advantage.

But the most important reason that people won't cheat is that leaders don't build useless, dangerous weapons. How long has it been possible to build biological weapons? At least since World War I. And how often have unscrupulous, immoral dictators waged war using biological weapons since then? That's right: not even once. And why is that? Because biological weapons are lousy weapons for fighting wars.

The proof of their lack of utility comes from the COVID pandemic. A new disease emerged and rapidly spread around the world. Despite sometimes extreme measures to contain it, it eventually reached basically everywhere on earth and killed eleven million people. The problem with biological weapons is that it is almost impossible to keep them from finding their way back to your own country and infecting your own people. They just aren't useful weapons.

When you stop and think about it, nuclear weapons have some of this same "poison blown on the winds" problem. Their effects are hard to keep contained. Their limited utility on battlefields or areas that are near battlefields is a function of their being "weapons of mass destruction."

Even unscrupulous, immoral dictators don't try to wage wars with weapons that are virtually useless (just as they don't guard their palaces with soldiers armed with dynamite). Unscrupulous, immoral leaders tend to seek out weapons that can help them win. If nuclear weapons are obsolete, the temptation to build them, which most people imagine is irresistible, will hardly exist. Because the weapons are so destructive, it will be necessary to set up stringent safeguards against someone building them. But it won't be very likely. People don't build weapons that aren't useful.

WORLD ELIMINATION FABLES

For some people, it's fairly easy to imagine a national campaign: nationwide campaigns of one sort or another happen fairly regularly. But it's not so easy to picture a worldwide campaign. So, to make it a little easier to see how the approach outlined here could spread around the world, let's have three imaginary stories about how the elimination movement worked once a national campaign got a government committed.

*   *   *

Imagine that it's twenty years after the treaty to eliminate nuclear weapons was signed and Mikaela, a high school student in the United States, and her mom are talking in the dining room by the food replicator.

MIKAELA

Mom: The big moment came in August 2025. President Tillman called a joint session of Congress and gave a speech that people still talk about. You can watch it on YouTube.

Mikaela: Mom, I keep telling you, no one watches YouTube anymore.

Mom: Well, it's still posted there, and maybe you can watch it there and just not tell your friends where you saw it. Your father and I watched it live on TV. Don't roll your eyes. I remember the hush that fell just before it started. The President had said that she was going to make a big announcement, and there was a lot of speculation about what it would be, but no one knew what she was going to say.

Then she started and said she had a dramatic policy change to announce. She paused and said, "From this day forward, it will be the policy of the United States that nuclear weapons are obsolete. These outmoded weapons are dangerous and barely useful at all, and from now on, we will treat them the way you would treat any kind of technology that has outlived its day." And it was as if everyone in Congress just caught their breath at the same moment. She went on to say that we wouldn't get rid of a single nuclear weapon … yet. But from now on, we would prepare for the day when they were gone. "As of today," she said, "nuclear weapons are in the garbage. They are, as far as the United States is concerned, trash waiting to get thrown away. But until there is a treaty requiring everyone to destroy their weapons at the same time, we will hold on to ours." We wouldn't spend any more money on them, we wouldn't upgrade them, or plan for more weapons in the future. And she said that from then on, the United States would be at the forefront of a movement to persuade every other country that these weapons were obsolete.

Mikaela: Did it work?

Mom: Well, it's interesting. Within a couple of months, Tillman was able to convince the Labour government in the United Kingdom to join in. They were spending about ten percent of their military budget on nuclear weapons, and they'd long ago lost faith that those weapons were anything more than symbols of grandeur. And a grassroots movement like the one in the United States had sprung up in the United Kingdom that brought real pressure to bear on politicians, which made the prime minister's decision much easier.

So, with the United Kingdom on board, Prime Minister Ingram and President Tillman called a special conference to "Reexamine Nuclear Weapons" and invited foreign ministers, diplomats, and heads of state from all over the world. Of course, because of the Treaty for the Prohibition of Nuclear Weapons—

Mikaela: —the TPNW—

Mom: —yes, the TPNW. Because the TPNW had prepared the way for all of the countries that don't have nuclear weapons, there was a lot of support from Latin American countries, African nations, and many of the Southeast Asian countries. All in all, there must have been forty countries that joined the coalition. Several of the European countries that had NATO nuclear weapons joined with the United States and United Kingdom—again, in part, because of really strong grassroots movements. But it turned out that the Southeast Asian countries were key. Their diplomats, according to later press reports—although it's still not clear exactly what happened—somehow convinced India to join the coalition. I personally think it was so easy to persuade India because I think they built nuclear weapons not for military reasons but to show that India was technologically sophisticated. They wanted the kind of respect due to the largest country in the world. They never really relied on nuclear weapons militarily. Their conventional military forces were stronger than Pakistan's, and the chances that China could fight its way through the highest mountains in the world in order to invade India were pretty limited. So, militarily, they didn't actually need the weapons.

Mikaela: Not to mention that their first political hero was a guy who would have been completely against nuclear weapons.

Mom: Yes, it is hard to imagine Gandhi endorsing nuclear weapons. So, India came on board. And then France joined. Their desire to be the most important voice in Europe was in jeopardy because Germany was so strongly against nuclear weapons, and it turned out that there had never been as much support for nuclear weapons among ordinary French people as there had been among their nuclear policy elite. So, with the United States, the United Kingdom, France, and India on board, the coalition started to really take shape. But probably the decisive moment came when China finally agreed to join. They had a long history of being ambivalent about the weapons at some level. And anyway, the Chinese prefer "soft power"—taking over economically—rather than conquering or destroying people.

So, when China came on board, there was suddenly a majority of the nuclear-armed states who thought the weapons were obsolete and wanted to get rid of them.

Mikaela: Which just left Russia, Israel, Pakistan, and North Korea.

Mom: Yes, and they were tough cases. Several of them really believed that their survival depended on nuclear weapons. But the United States put pressure on Israel and signed a stronger alliance and defense agreement with them, the Chinese reassured Pakistan and pressured North Korea, and in the end, it all came down to Russia. It took a couple more years, but even the Russians couldn't hang on when virtually the entire world was saying these weapons were obsolete. They've always had a chip on their shoulder about being called "backward," and the French in particular played on that with remarkable skill. It helped that the United States and China were rolling out new, smaller, smarter, swarming weapons that everyone was calling the weapons of the future. The contrast between these new, super-effective weapons and the blundering nature of nuclear weapons finally convinced the Russians. And once everyone agreed, the negotiations for the actual treaty only took about a year, the Treaty Eliminating Nuclear Weapons was signed—

Mikaela: The Ten-W.

Mom: —yes, the Ten-W—Ms. Hillebrand really has been teaching you guys about this—and the International Nuclear Weapons Monitoring Agency was set up, and each country, with elaborate U.N. monitoring, destroyed their weapons.

And the confidence that people got from the experience helped to solve a whole collection of other problems.

Mikaela: You always want to take credit for the CPT, but it was young people rising up all around the world that got it done.

Mom: Yes, young people were the spark that drove the Climate Protection Treaty to the front of the world's attention, but it was the success of the nuclear treaty that gave government leaders the confidence that such a big problem could be taken on and solved.

Mikaela: [Partly teasing, rolling her eyes] Whatever.

* * *

SOFIA

Sofia: I would like to thank Herr Richter for inviting me back to Französisches Gymnasium.[207] It is quite an experience to walk these halls again, and I would like to reassure whoever currently has locker A-342 that they shouldn't worry. Your stuff is safe. I tried to break into your locker, but they have changed the combination since my time here. [Laughter.] Despite what Herr Richter told you in his generous introduction [nods to Herr Richter], the part I played in the creation of the Vertrag über die Abschaffung von Atomwaffen [Nuclear Weapons Elimination Treaty] or VAA, was rather small. But I will be happy to talk to you about my part in the negotiations and what I learned from the experience.

But just briefly, let me remind you how the treaty came into existence. Because of the war in Ukraine in the 2020s, Germany became a leading voice in recommending the removal of nuclear weapons from European soil. This was possible only because Germany increased also its defense spending and enlarged its conventional forces at the same time under the political slogan "only useful weapons." This desire to have the nuclear bombs that the United States had stored in Germany removed eventually spread to the Netherlands, then Belgium, Italy, and eventually Turkey. The bombs were both dangerous and difficult to find a practical use for. They made Germany a target if a nuclear war were to happen.

One consequence of the debate spurred by Germany's determination to remove the weapons from its soil was an outpouring of German and, later, other European military white papers explaining in some detail why nuclear weapons were not very useful for the battlefield. These papers were, I believe, the key. Many people point to the grassroots movements that were also protesting and marching, but I think ultimately it was the expert's opinion that mattered. Eventually the government in Washington did withdraw the weapons, but that did not change their attitude immediately. They continued to proclaim that nuclear weapons were "the most important weapons," "essential for defense," and so on. But the withdrawal had an important impact on the public's perception of the weapons. It led to the widespread belief that nuclear weapons were no longer "necessary" or "ultimate" weapons. It permanently damaged their reputation.

The real change in the political fortunes of nuclear weapons came, however, with the British budget crisis of 2029. The Labour government then in power looked at their financial ledgers and decided that with more than ten percent of their military spending going to a tiny nuclear force they did not intend to ever use, they would scrap any future expansion or enhancements and, rather, spend the money on refurbishing hospitals. With their landslide victory the next year, Prime Minister Harris decided that the United Kingdom would join with Germany and the other countries that were arguing that nuclear weapons were obsolete.

With the added momentum that came with the United Kingdom joining the coalition promoting the idea that nuclear weapons are obsolete, and following the remarkable presidential elections of 2032, Chancellor Baerbock was able to persuade the new president, Honoré Durand, that France should join what was now being called the Preparatory Coalition. And in France, the grassroots movement was important. Durand had committed himself before the elections to work with Germany, and that fact was at least partly responsible for his landslide win. With Germany, the United Kingdom, and France on board, over the next two years the coalition expanded to include two states that had previously been under the U.S. nuclear umbrella (Japan and Australia) and several NATO states (Spain, Norway, Iceland, Canada, Croatia, and Greece).

And finally, after years of lobbying by the German government and other members of the Preparatory Coalition, in 2032, with the election of President Gutierrez, the United States grudgingly joined the coalition, declared that nuclear weapons were obsolete, and took up the effort of convincing other countries to join. And within four years after that, all the remaining nuclear-armed states had joined.

As I'm sure you all know, the treaty negotiations, which I had some role in and which are the main purpose of my talk here, took only eight months to complete; and in 2037, the VAA was signed, eliminating these dangerous and useless weapons.

So, about the negotiations. I was at home on a Saturday—gardening, actually—when my daughter Ulricka came out with my cell phone in her hand, saying there was a call from the Foreign Ministry.

* * *

LI QUÁNG BÓ

Edwards: Class, I'd like to introduce Li Quáng Bó to you. Professor Li is a distinguished scholar of modern Chinese history and is one of the chief historians of what we in the United States call the NuWET—the Nuclear Weapons Elimination Treaty. [Polite applause.]

Li: Thank you, Professor Edwards. It is a great pleasure to be here at Princeton University and to address your graduate seminar. I'm afraid I will have to run out in exactly forty-five minutes to make my flight to Beijing, so if you forgive me, I will dive right into the story of how this remarkable treaty came to be. While most Western countries believed the Mutual Assured Destruction theory as well as the idea that nuclear weapons were essential for defense, Chinese officials had never embraced nuclear weapons as enthusiastically as some in the West. Mao Zedong himself had called nuclear weapons a "paper tiger"—

Edwards: —Meaning they weren't militarily important—

Li: Yes. Meaning they weren't militarily important. And the Chinese nuclear arsenal, until the early 2020s, never grew to the proportions that U.S. and Russian arsenals did. The enthusiasm for nuclear weapons that had raised the number of nuclear weapons in the U.S. and Russian arsenals together to nearly seventy thousand weapons in the 1980s never took root in Chinese strategic thinking. In China, the number of long-range weapons, weapons that could reach the United States, for example, never rose above twenty-five for the first fifty-five years of the existence of Chinese nuclear forces. Although China eventually expanded its arsenal in the early 2020s, it never had as many missiles as either the United States or Russia.

Student: Excuse me, Professor Li. You're saying that Chinese leaders were always skeptical of the value of nuclear weapons?

Li: Yes. Chinese leaders were skeptical of the value of nuclear weapons. Historically, there had always been suspicion that the weapons had few practical uses. Nuclear weapons' status as the currency of power required China to have at least some of these weapons, but there were always doubts about the value of that currency. And we have always had a prejudice, so to speak, in favor of what you call "soft power" over hard power. So, when the Ukraine Nuclear Crisis erupted in the closing year of that war, China's leaders not only played a vital role in defusing the crisis; they also determined to undertake a sweeping reevaluate of Chinese nuclear weapons policy. The danger of that crisis, they decided, which brought the world again to the brink of nuclear war, could not be repeated. That sweeping review, undertaken over two years, eventually resulted in the historic Taiping Conference held in Hangzhou, which was attended by representatives of almost every country in the world and jointly hosted by the Chinese government and the United Nations. This landmark conference—it is widely agreed by historians of both East and West—was the beginning of the process that eventually led to the treaty. It was also—

Student: Would you say the conference was crucial because it accelerated the decline of the reputation of nuclear weapons?

Li [Annoyed by the Western habit of interrupting professors but not showing it]: Yes. Precisely. It appeared to be an example of the "tipping point" phenomenon put forward by your Malcolm Gladwell. There was an underlying dissatisfaction with nuclear weapons throughout the world which was quite strong, but which had not been openly expressed. The Taiping Conference appears to have unleashed this dissatisfaction—given it a voice, so to speak—and very quickly after the conference, countries began to proclaim Declarations of Obsolescence. In capital after capital over the next five years governments signed on to these Declarations.

Edwards: Professor Li, you're being modest. Isn't there now considerable evidence that Chinese diplomats actively promoted those declarations?

Li: [Carefully.] Professor Edwards, it is true that those rumors have often been repeated. But there is no hard evidence for them, and it is the position of the Chinese government that the movement to eliminate nuclear weapons was the result of a worldwide shift in attitudes, not diplomacy. The entire world deserves credit for those historic declarations. And of course, once the people's minds had been properly prepared, the ultimate United Nations treaty banning nuclear weapons was almost—I will not say, "nothing more than a formality"—but it was far easier than any previous nuclear weapons negotiation had been.

*　　*　　*

As these three fables demonstrate, with vigorous leadership, it is possible to shift world opinion—especially when the reality of the situation is so obvious. Negotiating nuclear weapons treaties is only hard now because of the overinflated reputation that nuclear weapons have. Dispel the myths, clear away the false theories, and insist that nuclear weapons are obsolete, and the task is much easier than we ever dared to hope.

So, there it is. The plan is not that complicated. It will be challenging and require determination and work, but it is based on well-known principles and political practices that have a long pedigree. And armed with the new perspective provided by the facts and arguments in this book, it can be done. It is, when you come down to it, remarkably straightforward.

Nuclear weapons are the currency of power. They symbolize power. What we are going to do is undermine faith in that currency. Just as monetary currencies can get overinflated when people are seized by a certain "irrational exuberance," so a currency of power can get overinflated. Our job is to use common sense, facts, and unrelenting realism to cause a market collapse in that currency, to dispel the myths that cause people to believe in nuclear weapons, and then stand back and watch the market crash. And once the transformation from currency of power to clumsy, blundering weapons is complete, we'll stand looking down at these rather ugly, dangerous, useless weapons and say to each other (with wonder in our voices), "How could people have ever thought these were worth keeping?"

THE FUTURE

We have lived for so long with nuclear weapons that less than two percent of the world's population now has any clear memory of what it was like to live in a world where they did not exist. It is hard to know what affect those years and years of imminent danger hanging over us have had. But it is useful, in trying to imagine the future, to know where you are starting from, to try to make a best guess at what their effect has been.

What impact have nuclear weapons had on societies around the globe? The world is complex, and it is difficult to tease out the impact of a single factor. Yet here in the United States, the effects are plain enough. Nuclear weapons have profoundly changed who we are.

In the United States, the taboo on talking about nuclear weapons is quite strong. But even though we say we never think about them, they haunt our dreams. When we imagine our future, nuclear weapons are there. Hollywood, the industry that delves into our subconscious and projects what it finds onto giant screens, unceasingly shows us futures that are devastated and dark. In the 1950s, less than ten movies were made that were predicated on some dark future catastrophe. By contrast, in the decade from 2010 to 2020, more than eighty movies about apocalypse and dystopia were made. Those movies sold an estimated 780 million or more tickets—

enough for every man, woman, and child to see a movie about a catastrophic future twice. Americans spent more than six billion dollars on these "entertainments."

And, if anything, these figures underestimate the extent to which we have watched fantasies about a future that has been destroyed. Streaming services like Netflix typically do not report viewing statistics. If the viewing habits of people sitting on their couches matches or even approximates those of people sitting in movie theaters, then the number of times people sat in dark rooms and absorbed these messages of despair could as much as double.

But movies are not the only place where dark dreams of a shattered future dwell. Books for young adults that tell tales of a dystopian future have been selling at a record pace. And before you say that they are just tales set in difficult circumstances to make the story more dramatic, listen to Jill Lapore, writing in a 2017 *The New Yorker* article titled, "A Golden Age for Dystopian Fiction." She points out that this wave of books about a dark future is not about triumphing over adversity. "Dystopia used to be a fiction of resistance; it's become a fiction of submission, the fiction of an untrusting, lonely, and sullen twenty-first century, the fiction of … helplessness and hopelessness."

The trend exists in television as well. There are more than ninety television shows, either current or in reruns, premised on a future that is devastated. But the place where these stories are truly metastasizing is video games—those staples of teenagers and twentysomethings. There are more than 350 games that present a picture of the future that is forlorn and broken. These games represent billions of hours of playing time. Fallout 4, for example, one of a series of games set in the near future in a United States reduced to ruins by a nuclear war, has been played more than two billion hours.

In the United States, we say we don't think about nuclear weapons or nuclear war—but that is simply a story we tell ourselves. Clearly, we do. What sort of impact does this constant imagining of a dark future have? What happens to a mind (or a heart) that regularly contemplates a future where everything is bleak?

The United States was, at one time, a nation characterized by optimism. De Tocqueville talked about a people perpetually looking toward the future. And that characterization was true for 150 years. When the Great Depression came in the 1930s, Americans didn't mutely hunker down, waiting for the Depression to pass. They decided to build their way out of it, and the signature project of that effort was the Grand Coulee Dam, the largest dam in the world at the time. Even facing hard times, they dreamed big. When the United States confronted the prospect of fighting two powerful adversaries on opposite sides of the globe, they never hesitated. Americans never had the slightest doubt that they would win.

Throughout its history, the United States has been defined by faith in its ability to make the future better. Sometimes that optimism has waxed; sometimes it has waned. But until recently, it was always there. Confidence was an abiding part of the American character.

Today feels different. There is now a major American television network that makes billions in profits by constantly peddling fear—fear of immigrants, fear that American values are being undermined, fear of government control. In business there is a kind of frantic insistence on profits now, this quarter, this month. The long-term has almost disappeared from popular business thinking. For individuals, spending on gambling has soared, as more and more people put their faith not in hard work and thrift but the luck of the draw. Deaths of despair, deaths from alcoholism, suicide, and drug addiction (especially opioid addiction) continue to climb. According to the National Institutes of Health, deaths from drug overdoses have increased fivefold—from twenty thousand to more than one hundred thousand per year—over the last twenty years.[208] What is it that people in the United States are so depressed about? And in politics—the arena where the collective future is hammered out—the divide between the two parties is so deep and the disagreements over the best way forward are so vehement that there appears to be little common ground left.

Far from seeming buoyant and optimistic, important parts of American society today reflect fearfulness, hopelessness, and self-destructiveness. And similar problems exist in other countries. The peculiar choice by the United Kingdom to withdraw from the European Union, for example, is a striking example of a self-inflicted wound. The foolish decision by Russia's leaders to try to conquer their way back to former glory shows a remarkable failure of judgment. The harsh treatment of Palestinians by Israel, accompanied by a slow shift from symbolizing holocaust survival to oppression seems inexplicable. There is a kind of gloom—a persistent lack of faith in the future, a strange lack of wisdom—in many places where nuclear weapons exist or where their influence extends.

So perhaps the way to think about living with nuclear weapons is that it is like living on death row, living with the threat of extinction always present. And some people's behavior today is reminiscent of the way some death row inmates live. They do their best to contract their world, set aside their hopes, and live only day to day. They seem to believe that to avoid bitterness and disappointment on the day you die, you must wall off your heart from any hopes or dreams.

This is the world that nuclear weapons have built. It is a difficult world, a troubling world, one in which the obstacles to real progress on important issues often seem insurmountable. So many people live lives where hope has been banished. So many act as if nothing matters. Talk about climate change and they respond, "Climate change? Whatever." They cannot seem to rouse themselves to care. They do their jobs, watch TV, and try not to think about larger issues.

If our present has been so affected by nuclear weapons and the insidious belief that they will always exist, then how would it change the future to have eliminated them? It would bring unimaginable change. In international affairs, in business, in politics, and in the way that ordinary people live, things would be radically different. And the change would be for the better.

At the global level, many nuclear weapons advocates believe that nuclear weapons provide the superstructure for the world order and, therefore, that a world without nuclear weapons would collapse into incoherence and anarchy. But the world survived in an orderly fashion for centuries before nuclear weapons existed, and it can be just as orderly today if we choose to make it so. At worst, the elimination of nuclear weapons would be like debunking any other symbol that was widely believed in. Because the reality is that they are not very good weapons and they are rather dangerous, their elimination would change nothing. Diplomatic rules, power dynamics, and the functioning of treaties would remain much the same. Alliances would be secured based not on the power of quasi-magical weapons to dissuade adversaries but on the same thing that secured them before nuclear weapons: shared interests, shared values, solemn oaths, and mutual regard.

Unraveling the current order where nuclear weapons play such a prominent role would bring many positive changes. There would, for instance, be more justice in the world. Nations with malign intentions, nations that invaded their neighbors, for example, would no longer be able to hide behind a shield of nuclear weapons. There would also be more realism. Small nations would no longer imagine that "ultimate" weapons could somehow keep them safe—in a way that no other technology had ever done before. Instead, they would have to fall back on the expedients that small nations have always relied on: alliances with strong nations and avoiding unnecessary conflicts with neighbors.

And there would be more cooperation. The experience of first negotiating and then putting such an important treaty into effect would strengthen people's confidence that they could tackle other problems that require cooperation. There would be a greater emphasis on human interactions in the construction of international relations rather than on the supposed power of technology to set boundaries and rules.

In business, the elimination of nuclear weapons would make the case for long-term thinking and investing more persuasive. It makes little sense to think about putting money into projects that may take years to mature if you're barreling along a precipice and your car might plunge over the edge at any moment. On the other hand, if the road before you stretches out over a vast, flat plain—vanishing into the distance—pondering long-term investments would seem quite sensible.

But the impact on politics would be even more profound. In a future without nuclear weapons, it would be possible to have larger agendas. Projects could be bigger and last longer. The constant, frenetic rush to accomplish everything today, to secure outsized results now, to ignore problems and undertakings which would require a longer time frame of effort, would fade into memory. When people feel there is a good chance the future will never come, it is hard to inspire change. When that feeling is replaced by a sense of certainty, it will be possible to once again make big plans, to imagine in decades rather than days.

But the greatest change would be in how we feel. It has been so long since the sunny days of the past that we have forgotten what it felt like to live without dark storm clouds constantly looming. A future without these weapons would revive a feeling of optimism, a sense of confidence, a faith that the future might actually come. Daring projects—like sending astronauts to the moon in eight short years—would sound inviting again. A future without the constant danger of catastrophic war would restore, in other words, our sense of hope.

Hope is the essential engine of all human accomplishment. It is the thing we need to survive. Hope can sometimes seem like a small thing. But it is what allows people in danger, people pressed hard by circumstances to continue to fight for existence. It is, next to love, one of the fundamental components of human nature.

Challenges that seem impossible today would suddenly seem within our grasp if our list of recent accomplishments had a check mark next to the words "eliminated nuclear weapons." World problems would seem smaller, less formidable. We have lost a certain amount of faith in our ability to do great things. Eliminating nuclear weapons would go a long way toward restoring that confidence.

Restoring hope would open new doors of change. William James said that character is not something you are born with; it is something that is earned, something that you build with each choice you make. Each decision is affected by the previous one, and each choice affects the one that follows—adding or subtracting to your character. Choosing to eliminate nuclear weapons would change the character of who we, collectively, are. It would change the trajectory of our future. Just as a tiny change in the course of a ballistic missile early in its flight leads, with every second it flies, to a greater change in its eventual landing place, so the decision to eliminate nuclear weapons now would change the arc of human civilization for centuries to come, with increasing importance as the years pass by. Choosing a future without nuclear weapons is the first step in building the institutions that can, over time, make us into better human beings.

There is no reason that human beings can't return to a time when confidence was the norm. There is no reason we can't return to a sense of hopefulness. We can, in a sense, break out of our cells. We can reach backward to an optimism that we have largely forgotten, and we can reach forward to a new, more evolved way of living. We can become not only as good as we were in the past but better than we have ever been.

A future without nuclear weapons would be a time of wider horizons. It would be larger, with far more possibilities. What I hate most about nuclear weapons is not the fear they inject into our hearts but the way they limit our imaginations. Nuclear weapons contract our horizons. There is much that we have yet to do, that we can do. Are we so hemmed in by fear and anxiety that we really believe our world cannot be preserved? I believe the human spirit is bigger and bolder than that. Life is more vibrant; it is more filled with excitement than the numb insistence that nothing can be done.

Wouldn't it be better to shrug off these weapons with their false sense of invulnerability, their dark fever dream of power and control, and cast them to the side of the road? Wouldn't it be better to head off toward a future washed clean? We have a duty to the people of tomorrow to do better than this. We are required by those who have yet to be born to find a path forward. We have lived so long cramped and hunched over, waiting for the expected blow, isn't it time to stand up, breathe deep, take risks, and go boldly toward some worthy goal? No individual and no nation achieves greatness trapped behind walls. Cowering in this prison of "security" is no sort of living.

The future is calling us. It is a future both bold and pragmatic. Now is the hour. No more excuses.

It is possible.

[204] Malcolm Gladwell, *The Tipping Point: How Little Things Can Make a Big Difference* (New York: Little, Brown, 2000).
[205] Nuclear Weapons Ban Monitor, 2022. https://banmonitor.org/news/nuclear-weapons-ban-monitor-2022-is-here.
[206] Ward Wilson, "Should public opinion polls influence America's nuclear policy?" The Hill, November 18, 2021, https://thehill.com/opinion/national-security/581404-should-public-opinion-polls-influence-americas-nuclear-policy/ (accessed May 25, 2023).
[207] One of the best secondary schools in Berlin.
[208] "Drug Overdose Death Rates," National Institute on Drug Abuse (NIDA), https://nida.nih.gov/research-topics/trends-statistics/overdose-death-rates#:~:text=More%20than%20106%2C000%20persons%20in,drugs%20from%201999%20to%202021. (accessed May 25, 2023).

APPENDIX

There are any number of worthy groups you can join. I'm kind of partial to RealistRevolt because I founded it and it is based on the arguments you've been reading in this book.

You can visit us at www.realistrevolt.org and find materials on organizing, more in-depth writing about these weapons, answers to your questions about how to carry this forward, and a community of people who believe that these weapons are too dangerous to continue to exist. RealistRevolt wants to help you explain the situation to others, build the power we need, and then—together—eliminate the danger once and for all.

There are, of course, other groups, with different approaches that might appeal more to you. Here is a list and brief summary of each one's mission:

BACK FROM THE BRINK
Back from the Brink is a U.S.-based grassroots coalition of individuals, organizations and elected officials working together toward a world free of nuclear weapons and advocating for common sense nuclear weapons policies to secure a safer, more just future. We call on the United States to lead a global effort to prevent nuclear war by: 1) Actively pursuing a verifiable agreement among nuclear-armed states to eliminate their nuclear arsenals. 2) Renouncing the option of using nuclear weapons first. 3) Ending the sole, unchecked authority of any U.S. President to launch a nuclear attack. 4) Taking U.S. nuclear weapons off hair-trigger alert. 5) Cancelling the plan to replace the entire U.S. nuclear arsenal with enhanced weapons. https://preventnuclearwar.org/

BAN MONITOR

The Ban Monitor tracks progress towards a world without nuclear weapons and highlights activities that stand between the international community and the fulfillment of the United Nations' long-standing goal of the elimination of nuclear weapons. In measuring progress, the Ban Monitor uses the TPNW as the primary yardstick, because this Treaty codifies norms and actions that are needed to create and maintain a world free of nuclear weapons. https://banmonitor.org

BASIC

We promote meaningful dialogue amongst governments and experts in order to build international trust, reduce nuclear risks, and advance disarmament. We envision a world that uses cooperative measures, rather than the threat or use of force, to achieve peace and security. This world will be achieved by taking steps that promote mutual security at the international, regional, national, and individual levels, and sustained through resilient international norms and law. https://basicint.org/

CENTER FOR ARMS CONTROL AND NONPROLIFERATION

The Center for Arms Control and Nonproliferation, based in Washington, D.C., is a national nonpartisan 501c(3) non-profit seeking to reduce nuclear weapons arsenals, halt the spread of nuclear weapons and minimize the risk of war by educating the public and policymakers. We focus heavily on education efforts to Members of Congress, Congressional staffers, the media and general public via public and private outreach. https://armscontrolcenter.org/

CENTER FOR THE STUDY OF EXISTENTIAL RISK

The Center for the Study of Existential Risk is an interdisciplinary research center within the University of Cambridge who study existential risks, develop collaborative strategies to reduce them, and foster a global community of academics, technologists, and policymakers working to safeguard humanity. Our research focuses on biological risks, environmental risks, risks from artificial intelligence, and how to manage extreme technological risk in general. CSER's work on nuclear weapons includes the global consequences of nuclear weapon use (such as climate impacts on human societies, cascade, and social effects). We also look at the management of uncertainty with regards to nuclear weapons, doubts around the stability of nuclear deterrence, and the impacts of emerging disruptive technologies. https://www.cser.ac.uk

APPENDIX

COUNCIL FOR A LIVABLE WORLD

Council for a Livable World, based in Washington, D.C., is a national nonpartisan 501c(4) nonprofit founded in 1962 by Manhattan Project nuclear physicist Leo Szilard. Szilard's vision was for an organization that would both educate decision-makers and the public on nuclear issues, and raise money for Congressional candidates seeking to reduce the nuclear threat. Since our founding, we have helped elect nearly 400 members of Congress who share our vision for a more livable world.
https://livableworld.org/

FEDERATION OF AMERICAN SCIENTISTS: NUCLEAR INFORMATION PROJECT

The Nuclear Information Project, called "one of the most widely sourced resources for nuclear warhead counts" by *The Washington Post*, uses open sources such as official documents, testimonies, previously undisclosed information obtained through the Freedom of Information Act, as well as independent analysis of commercial satellite imagery as the basis for developing the best available unclassified estimates of the status and trends of nuclear weapons worldwide. The Project also conducts analysis of the role of nuclear weapons and provides recommendations for responsibly reducing the numbers and role of nuclear weapons.
https://fas.org/issue/nuclear-weapons/

FUTURE OF LIFE INSTITUTE

How certain technologies are developed and used has far-reaching consequences for all life on earth. This is currently the case for artificial intelligence, biotechnologies and nuclear technology. If properly managed, these technologies could transform the world in a way that makes life substantially better, both for the people alive today and for all the people who have yet to be born. They could be used to treat and eradicate diseases, strengthen democratic processes, and mitigate—or even halt—climate change. If improperly managed, they could do the opposite. They could produce catastrophic events that bring humanity to its knees, perhaps even pushing us to the brink of extinction. The Future of Life Institute's mission is to steer transformative technologies away from extreme, large-scale risks and towards benefiting life. https://futureoflife.org/

GLOBAL ZERO

We envision a convergence of better futures, far beyond the limits of the belief that nuclear weapons somehow keep us safe. We can see they do not. Imagine a future where stability is not precariously balanced on the threat of mass destruction; where safety for some no longer requires vulnerability for others; where justice and equity are experienced by communities most impacted by nuclear harm; and where international cooperation in the face of common threats allows us to finally address the many other urgent challenges competing for attention. Climate. Public Health. Racial Justice. Global Zero is committed to operationalizing our values. Everything we do—from communications and movement building, to governance and professional development—is informed by our commitment to cultivate imagination and empathy, and take actions that call on courage and build trust, all in the fierce pursuit of justice. https://www.globalzero.org/

INTERNATIONAL CAMPAIGN TO ABOLISH NUCLEAR WEAPONS (ICAN)

ICAN is a broad, inclusive campaign, focused on mobilizing civil society around the world to support the specific objective of prohibiting and eliminating nuclear weapons. The ICAN international structure consists of partner organizations, an international steering group and an international staff team. ICAN won the Nobel Peace Prize in 2017. https://www.icanw.org/

INTERNATIONAL COMMITTEE FOR THE RED CROSS (ICRC)

The work of the ICRC is based on the Geneva Conventions of 1949, their Additional Protocols, its Statutes—and those of the International Red Cross and Red Crescent Movement—and the resolutions of the International Conferences of the Red Cross and Red Crescent. The ICRC is an independent, neutral organization ensuring humanitarian protection and assistance for victims of armed conflict and other situations of violence. It takes action in response to emergencies and at the same time promotes respect for international humanitarian law and its implementation in national law. https://www.icrc.org/en

APPENDIX

INTERNATIONAL PHYSICIANS FOR THE PREVENTION OF NUCLEAR WAR

IPPNW is the only international medical organization dedicated to the abolition of nuclear weapons. Founded by U.S. and Russian physicians in 1980, IPPNW is credited with raising public awareness about the devastating effects of nuclear weapons and with persuading American and Soviet leaders that the Cold War nuclear arms race was jeopardizing the survival of the entire world. IPPNW received the 1985 Nobel Peace Prize in recognition of this accomplishment.

Today IPPNW mobilizes doctors, medical students, and concerned citizens in over 60 countries in the service of a broader war prevention mission. The International Campaign to Abolish Nuclear Weapons (ICAN) and Aiming For Prevention, IPPNW's campaign to prevent armed violence worldwide, bring the expertise and compassion of doctors to bear on the whole human tragedy of armed conflict. https://www.ippnw.org

NSQUARE

N Square brings new energy and a fresh approach to tackling one of the world's wickedest problems: the deepening threat posed by nuclear weapons. The nuclear landscape is a jumble of many overlapping challenges, from political dilemmas to technological puzzles to conflicts between belief systems. We recognize these complexities without being daunted by them, employing a range of tactics to create conditions that enable new solutions, ideas, and partnerships. https://nsquare.org/

NUCLEAR BAN US

NuclearBan.US is a 501c (4) non-profit organization committed to achieving a world free from nuclear weapons by persuading the US to sign, ratify, and implement the Treaty on the Prohibition of Nuclear Weapons (TPNW). NuclearBan.US is a partner of ICAN, which won the 2017 Nobel Peace Prize for its part in negotiating this historic treaty (also known as the 'Nuclear Ban Treaty'). https://www.nuclearban.us/

NUCLEAR THREAT INITIATIVE

NTI's vision is of a world safe from preventable global catastrophe. Its mission is to transform global security by driving systemic solutions to nuclear and biological threats imperiling humanity. NTI collaborates with governments and organizations to raise awareness, advocate, and implement creative solutions. As an independent and trusted partner, we transcend traditional thinking and stimulate new ways to address urgent threats. https://www.nti.org/

MASSACHUSETTS PEACE ACTION (MAPA)

Massachusetts Peace Action is a nonpartisan, nonprofit organization working to develop the sustained political power to foster a more just and peaceful U.S. foreign policy. Through grassroots organizing, policy advocacy, and community education, we promote human rights and global cooperation, seek an end to war and the spread of nuclear weapons, and support budget priorities that redirect excessive military spending to meeting human and environmental needs in our communities.

We are an affiliate of Peace Action, the nation's largest grassroots peace and disarmament membership organization, with more than 100,000 members and 30 chapters across the country.

We share a vision of world peace in which: the menace of nuclear weapons has forever been erased from our planet; war has been abolished as a method of solving conflicts; all human beings are assured the wherewithal to live in health and dignity; no one is denied the opportunity to participate in making decisions that affect the common good. https://masspeaceaction.org/our-issueshttps://masspeaceaction.org/nuclear-disarmament/

ONE EARTH FUTURE

Open Nuclear Network (ONN) is a project of One Earth Future. The goal of Open Nuclear Network is to reduce the risk that nuclear weapons are used in response to error, uncertainty or misdirection, particularly in the context of escalating conflict, using innovation, inclusion and dialogue supported by open-source data. Today, more than ever before, nuclear weapons are at risk of being used due to human or technical error, uncertainty about adversaries' capabilities or intentions or intentional misdirection, particularly in the context of escalating conflict. The mission of ONN is to reduce the risk of the inadvertent use of nuclear weapons, using innovation, inclusion and dialogue supported by open-source data.

ONN works on nuclear risk reduction in two complementary ways: A team of analysts collects data and produces informational briefs, leveraging publicly available data and state-of-the-art technology. In tandem, ONN deploys a network of trusted third parties for data-enabled efforts to improve dialogue among world leaders and decision makers and promote nuclear de-escalation. https://opennuclear.org/

PAX — DON'T BANK ON THE BOMB

Don't Bank on the Bomb is the only regularly published source of information on the private companies involved in the production of nuclear weapons and their financiers. The report examines contracts for the production of key components of nuclear weapons and their specifically designed delivery systems. It provides information on the financial institutions seeking to profit from these producing companies. The report also profiles those institutions and others that limit or prohibit any financial engagement with companies associated with the production of nuclear weapons. www.dontbankonthebomb.com/

PEACE ACTION

Peace Action works for smarter approaches to global problems. If we want to address problems like war, the nuclear threat, poverty, climate change, and terrorism—the U.S. needs to work together, cooperatively, with other nations. We also need to overcome the partisan politics and divisive rhetoric that often drown out alternatives to war. By getting people and communities around the country involved, we build the political will needed to break through that deadlock. Our success comes from empowering people to engage in foreign policy issues like no other organization. https://www.peaceaction.org/

PHYSICIANS FOR SOCIAL RESPONSIBILITY

Guided by the values and expertise of medicine and public health, we work to protect human life from the gravest threats to health and survival. PSR mobilizes physicians and health professionals to advocate for climate solutions and a nuclear-weapons-free world. https://psr.org/

PROGRAM ON SCIENCE AND GLOBAL SECURITY

Building a safer, more peaceful world. Princeton University's Program on Science and Global Security (SGS), based in the School of Public and International Affairs, conducts scientific, technical and policy research, analysis and outreach to advance national and international policies for a safer and more peaceful world.

The Program was founded in 1974 by physicists in Princeton's School of Engineering and Applied Science as part of Center for Energy and Environmental Studies and moved to Princeton's School of Public and International Affairs in 2001. Harold Feiveson and Frank von Hippel founded and co-directed the Program from 1974 to 2006. From 2006 to 2016, the Program was directed by Christopher Chyba. Since 2016, SGS has been directed by Alexander Glaser and Zia Mian.

Throughout its history, SGS has worked on nuclear arms control, nonproliferation, and disarmament to reduce the dangers from nuclear weapons and nuclear power. It is one of the oldest and most highly regarded academic programs focused on technical and policy studies on nuclear issues in the world. In the past decade it also has advanced policy on biosecurity issues. SGS engages effectively and creatively with these long-standing policy issues that remain to resolved and tracks emerging challenges from disruptive technologies with the potential to transform global security. These include new biotechnologies, information and communications technologies, autonomous weapons, artificial intelligence, and space-based systems. https://sgs.princeton.edu/

REALISTREVOLT

The fundamental message of RealistRevolt is one of realism and hope. It is not only possible, but also imperative to eliminate nuclear weapons. We believe this not because we are idealists, but because we are realists. It is not realistic to believe that a miraculous technology can guarantee our safety. Realists know that there are no guarantees in life. It is not realism to rely on a weapon to prop up the world order. Realists know that world order is built on law and rules that are fairly enforced. It is not realism to believe that our prosperity is sustained by the threat of annihilation. Despite their constant clamor, apologists for nuclear weapons are not realists, and the rationale for keeping these weapons is not realistic. We are looking for partners who are serious and determined; who are realists but at the same time who know that change is possible; partners who have the capacity to hope.

APPENDIX

ROTARY WORLD
A worldwide Rotary club focused on nuclear abolition.
RotarysatelliteWorld@gmail.com

THE WILLIAM J. PERRY PROJECT
Founded by U.S. Secretary of Defense William Perry to change the conversation around nuclear weapons, and to educate the public on the nuclear threat in the 21st century. https://www.wjperryproject.org/

UNION OF CONCERNED SCIENTISTS
The Union of Concerned Scientists is a national nonprofit organization founded more than 50 years ago by scientists and students at the Massachusetts Institute of Technology. Our mission: to use rigorous, independent science to solve our planet's most pressing problems. We are a group of nearly 250 scientists, analysts, policy experts, and strategic communicators dedicated to that purpose. We combat climate change and seek to alleviate harm caused by the heat, sea level rise, and other consequences of runaway emissions, we strive to develop sustainable ways to feed, power, and transport ourselves, we work to reduce the existential threat of nuclear war, we fight back when powerful corporations or special interests mislead the public on science, and we ensure our solutions advance racial and economic equity. https://www.ucsusa.org/

WIN WITHOUT WAR
Win Without War is a diverse network of activists and organizations working for a more peaceful, progressive U.S. foreign policy. We believe that by democratizing U.S. foreign policy and providing progressive alternatives, we can achieve more peaceful, just, and commonsense policies that ensure that all people—regardless of race, nationality, gender, religion, or economic status—can find and take advantage of opportunity equally and feel secure. https://winwithoutwar.org

INDEX

B-29 Bomber, 144, 190, 191
Battles
   Agincourt in 1415, 69
   Crécy in 1346, 69
   Dien Bien Pshu in 1954, 116, 117, 121, 122, 140
   Heraclea in 280 BC, 154
   Vicksburg in 1863, 131
   Zama in 202 BC, 88
Bigness Quotient, 83, 84

Castro, Fidel, 152, 182, 183, 203
Cheney, Dick, 118
Churchill, Winston, 30, 47, 95, 104, 153, 177
Cuban Missile Crisis, 2, 6, 137, 141, 168, 170, 172, 182, 187, 195, 196, 197, 202, 203, 206

DeGroot, Gerard, 81, 82, 89, 91, 95, 110
Disinvention, iii, 71, 72, 73, 74, 76, 78
Douhet, Giulio, 60, 64
Dreadnought, 38, 39, 40, 42, 44, 87, 88
Dyson, Freeman, vii, 129, 130, 140, 141, 180, 187, 202, 206

Einstein, Albert, 65, 79, 96
Eisenhower, Dwight, 63, 115, 116, 118, 124, 125, 129, 160, 179
Elephants, 69, 87, 88

Framing, 16, 22, 23, 28, 71, 73, 98

Gaddis, John Lewis, 30, 34, 110, 118, 140, 141, 172, 196, 206
Gorbachev, Mikhail, 213

Haldeman, H. R., 179, 187
Hirohito, 107, 111, 112
Hiroshima, 4, 9, 12, 17, 23, 52, 94, 95, 96, 97, 99, 100, 101, 102, 103, 104, 107, 115, 122, 123, 124
Hitler, Adolf, 105, 120, 121, 224

Kahn, Herman, 8, 96
Kennedy, John F., 30, 135, 137, 168, 169, 195, 196, 203
Kennedy, Robert F., 7, 8, 33, 202
Khrushchev, Nikita, 137, 152, 168, 169, 172, 182, 183, 203
Khwarazmian Empire, 147, 148

LeMay, Curtis, 28, 144

MacArthur, Douglas, 115, 135
McCarthy, Joseph, 56, 57, 63
McNamara, Robert, 150, 168, 199, 203, 204
Meir, Golda, 119, 120

Nagasaki, 9, 17, 23, 52, 83, 94, 95, 97, 99, 101, 103, 124
Napoleon, 145, 146
Nixon, Richard, 129, 178, 179, 180, 187, 192

Oppenheimer, J. Robert, 34, 50, 52, 63, 89

Paris Gun, 75, 76, 211
Petrov, Stanislav, 200
Powell, Colin, 117, 118, 119
Ptolemy's Giant Warship, 74, 75, 76, 211
Pyrrhus of Epirus, 154

Reagan, Ronald, 21, 82, 129, 213
Roosevelt, Franklin D., 93
Roosevelt, Theodore, 39, 40

Schelling, Thomas, 8, 150, 176, 177, 178, 187, 189
Schlesinger, James, 27, 133, 134, 141
Stalin, Joseph, 25, 26, 34, 105, 111, 177, 190
Stimson, Henry, 79, 93, 94, 97, 98, 104, 110

Taiwan Straits, 118, 124

TPNW
    The Treaty on the Prohibition of Nuclear, 22

Truman, Harry, 52, 79, 93, 95, 100, 110, 111, 114, 136, 137, 141, 156

Wars
    Cold War, 11, 17, 22, 30, 34, 56, 57, 58, 61, 62, 69, 79, 110, 126, 139, 140, 141, 172, 189, 196, 211, 243
    Falklands, 68, 187, 193, 196
    Gulf War, 117, 118, 119
    Korean, 67, 179, 187, 206
    Middle East War, 68, 119, 120, 186, 187, 192, 193, 197
    Vietnam, 67, 116, 117, 140, 163, 164, 178, 179, 180
    WW I, 14, 41, 42, 43, 44, 59, 75, 143, 201, 224
    WW II, 3, 12, 28, 42, 43, 44, 45, 54, 56, 59, 60, 67, 69, 90, 93, 96, 97, 105, 106, 111, 113, 114, 115, 116, 121, 124, 128, 129, 131, 135, 144, 146, 150, 160, 190, 224
Wilhelm II, Kaiser, 40, 41, 44, 47

ABOUT THE AUTHOR

Ward Hayes Wilson is the executive director of RealistRevolt and the author of *Five Myths About Nuclear Weapons*, which was endorsed by Pulitzer Prize-winning historians of nuclear weapons, military leaders, and Nobel Peace Prize laureates. Wilson has published widely in both popular and academic journals including *The New York Times*, *The Boston Globe*, *The Wall Street Journal*, *The Nation*, *The Chicago Tribune*, *The Bulletin of the Atomic Scientists*, *Nonproliferation Review*, *Joint Force Quarterly*, *Parameters*, *Revue de Défense Nationale*, *Survival*, *Foreign Policy*, *International Security*, *Ethics and International Affairs*, *The Diplomat*, and others. Having appeared on national and international television, radio, and podcasts, Wilson is a sought-after public speaker who has presented in government, academic, and popular settings in 23 countries on six continents.

Printed in Great Britain
by Amazon

4c2aaf46-4162-4a72-825b-e2689397a471R01